Lecture Notes of the Institute for Computer Sciences, Social Informatics and Telecommunications Engineering   52

Prashant Pillai Rajeev Shorey
Erina Ferro (Eds.)

Personal
Satellite Services

4th International ICST Conference, PSATS 2012
Bradford, UK, March 22-23, 2012
Revised Selected Papers

 Springer

Volume Editors

Prashant Pillai
University of Bradford
School of Engineering, Design and Technology
Richmond Road
Bradford, BD7 1DP, UK
E-mail: p.pillai@bradford.ac.uk

Rajeev Shorey
NIIT University
NH-8, Delhi-Jaipur Highway
Neemrana, 301705 District Alwar, Rajasthan, India
E-mail: rajeev.shorey@niituniversity.in

Erina Ferro
ISTI Institute
CNR Research Area
Via G. Moruzzi 1
56124 Pisa, Italy
E-mail: erina.ferro@isti.cnr.it

ISSN 1867-8211   e-ISSN 1867-822X
ISBN 978-3-642-36786-1   e-ISBN 978-3-642-36787-8
DOI 10.1007/978-3-642-36787-8
Springer Heidelberg Dordrecht London New York

Library of Congress Control Number: 2013931767

CR Subject Classification (1998): C.2.1-2, C.2.6, H.4.3, J.1, J.2, C.4, K.6.5

Typesetting: Camera-ready by author, data conversion by Scientific Publishing Services, Chennai, India

Printed on acid-free paper

Springer is part of Springer Science+Business Media (www.springer.com)

Preface

The 4th International Conference on Personal Satellite Services (PSATS) was held at the Norcroft Centre at the University of Bradford, UK, during March 22–23, 2012.

The next-generation satellite services will cater to the demands of personal services by bringing the satellite terminals directly to the hands of the user, hence providing satellite personal services directly to the user. Technological advances in satellite communications have made it possible to bring such value-added satellite services directly to the user by reducing the overall cost as well as addressing several technological challenges. A new category defined as personal satellite services (PSATS) extends satellite services directly to users for personal services such as communications, multimedia, and location identification. PSATS 2012 was a two-day event that explored such techniques and provided a platform for discussion between academic and industrial researchers, practitioners, and students interested in future techniques relating to satellite communications, networking, technology, systems, and applications.

PSATS 2012 boasts of two outstanding keynote speakers. We were fortunate to have Cosimo La Rocca, Advisor to the President of the Italian Space Agency ASI, Italy, and Surendra Pal, Professor and Senior Adviser, Satellite Navigation, Indian Space Research Organisation, India, as our keynote speakers. PSATS 2012 also included two tutorials by eminent experts in the area. The first tutorial titled "Quality of Service in Terrestrial Wireless and Satellite Networks for Mobile Personal Services" was offered by Giovanni Giambene, University of Siena, Italy. The second tutorial titled "Resource Management in Cloud Computing" was offered by Nicola Tonellotto, Information Science and Technologies Institute, National Research Council of Italy.

PSATS 2012 included four technical sessions consisting of 22 high-quality regular papers and one poster session with five posters on advanced topics in satellite communications. The four technical sessions included: (1) Radio Resource Management, (2) Spectrum/Interference Management and Antenna Design, (3) Mobility and Security, and (4) Protocol Performance in Satellite Networks. PSATS 2012 also included a panel discussion session on the topic: "The role of satellites in future personal and vehicular communications." This session provided a platform for sharing the views of the satellite operators, satellite research centers, SMEs, and universities on the future role of satellite communications in our everyday life.

The conference had close to 60 participants both from the industrial and the academic sectors from various parts of the world such as India, France, Germany, Greece, UK, and Italy. The conference provided tea and coffee breaks as well as lunch at the conference centre for all the participants. A Gala Dinner event was also organized on the first day of the conference. The quality of the venue, services provided by the center staff, and especially the quality of food and drinks were all highly spoken about during the conference by the participants and the Organizing Committee as well.

Last but not least, we would like to thank the Organizing Committee members, Session Chairs, Technical Program Committee members, all the authors and speakers for their technical contributions and the attendees for their participation. Also, on behalf of the Organizing Committee and the Steering Committee of PSATS, we would like to thank our sponsors, ICST, University of Bradford, NIIT University, ASI, and Heaton Education for their generous financial support and CREATE-NET, Eutelsat, and EAI for their extended support in making this event a successful one.

Prashant Pillai
Rajeev Shorey
Erina Ferro

Organization

General Co-chairs

Prashant Pillai   University of Bradford, UK
Rajeev Shorey   NIIT University, India

Technical Program Co-chairs

Erina Ferro   ISTI-CNR, Pisa, Italy
Neeli R. Prasad   Aalborg University, Denmark

Steering Committee

Imrich Chlamtac   Create-Net, Italy
Kandeepan Sithamparanathan RMIT University, Australia
Agnelli Stefano   ESOA/Eutelsat, France
Mario Marchese   University of Genoa, Italy

Advisory Committee

Giovanni Giambene   University of Siena, Italy
Fun Hu   University of Bradford, UK
Vinod Kumar   Alcatel-Lucent, France

Industrial Chair

Paul Febvre   Inmarsat, UK

Publicity Chair

Tomaso De Cola   German Aerospace Center DLR, Germany

Exhibition Chair

Ana Yun Garcia   Thales Alenia Space, Spain

Publications Chair

Constantinos T. Angelis   TEIE, Greece

Tutorial Chair

Alberto Gotta   ISTI-CNR, Pisa, Italy

Local Organizing Chairs

Yongqiang Cheng   University of Bradford, UK
Kai Xu   University of Bradford, UK

Conference Coordinators

Aza Swedin   EAI, Italy

Website Chair

Konstantinos Kotsopoulos   University of Bradford, UK

Technical Program Commitee

Roberto Di Pietro   University of Rome, Italy
Gorry Fairhurst   University of Aberdeen, UK
Anand Prasad   NEC Corporation, Japan
Carlo Caini   University of Bologna, Italy
Matteo Berioli   German Aerospace Center (DLR), Germany
Alberto Gotta   ISTI-CNR, Italy
Raffaello Secchi   University of Aberdeen, UK
Laurent Franck   Télécom Bretagne, France
Franco Davoli   CNIT, Italy
Paolo Barsocchi   ISTI-CNR, Italy
Petia Todorova   Fraunhofer Institut FOKUS, Germany
Carles Fernandez-Prades   CTTC, Spain
Gabriele Oligeri   ISTI-CNR, Italy
Ajay Kulkarni   Cisco Systems, USA
Marco Cello   University of Genoa, Italy
Alexey Vinel   SPIIRAS, Russia
Fabio Dovis   Politecnico di Torino, Italy
Igor Kotenko   SPIIRAS, Russia
Alban Duverdier   CNES, France
Cesare Roseti   University of Rome, Italy
Anton Donner   German Aerospace Center (DLR), Germany
Ernestina Cianca   University of Rome Tor Vergata, Italy
Harald Skinnemoen   AnsuR Technologies, Norway
Bernhard Collini-Nocker   University of Salzburg, Austria

Table of Contents

Mobility and Security

Protocol Performance in Satellite Networks

Poster Session on Satellite Communications

rtPS Scheduling with QoE Metrics in Joint WiMAX/Satellite Networks

Anastasia Lygizou, Spyros Xergias, and Nikos Passas

National and Kapodistrian University of Athens, Dept. of Informatics and Telecommunications,
Panepistimiopolis Ilissia, 15784 Athens, Greece
{lygizou,xergias,passas}@di.uoa.gr

Abstract. This paper improves a previously proposed scheduling algorithm that is responsible to share the allocated capacity to the uplink traffic of an integrated satellite and WiMAX network. The target of this improvement is to schedule traffic of real time connections based on Quality of Experience (QoE) metrics. After a bibliographic search on QoE metrics, the FC-MDI (Frame Classification-Media Delivery Index) metric is chosen to be used in the proposed algorithm for the scheduling of real time connections. Two versions of the algorithm are proposed and evaluated. Simulation results show that the proposed algorithm considerably improves the QoE and the mean delay of the real time connections.

Keywords: DVB-RCS, WiMAX, QoE, rtPS, scheduling.

1   Introduction

IEEE 802.16 [1] is a standard that aims at filling the gap between local and wide area networks, by introducing an advanced system for metropolitan environments. In this system, also known as WiMAX, both point-to-multipoint (cellular) and mesh mode configurations can be supported, while node mobility is also covered by amendment 802.16e [2]. One of the main advantages of the standard is the large degree of flexibility it provides by supporting a wide range of traffic classes with different characteristics and quality of service (QoS) requirements. This is attained through a large set of parameters that allow users to describe in detail their traffic profiles and service needs. On the other hand, Digital Video Broadcasting – Return Channel Satellite (DVB-RCS) [3] is an open standard for bi-directional transmission of digital data over the satellite network. DVB-RCS is a fully mature open, satellite communication standard with highly efficient bandwidth management, making it a cost-efficient alternative in many cases. It mainly describes the uplink direction of a satellite network, providing advanced QoS capabilities for requesting and acquiring capacity for demanding services.

The advantage of combining the two technologies is that a satellite network can be used for interconnecting WiMAX islands with the Internet and avoiding layout of expensive backbone infrastructures. This can provide reliable solution, especially in rural areas or locations affected by environmental factors, e.g. islands, mountains, etc. However, a satellite network experiences large round trip delays that can deteriorate

P. Pillai, R. Shorey, and E. Ferro (Eds.): PSATS 2012, LNICST 52, pp. 1–8, 2013.
© Institute for Computer Sciences, Social Informatics and Telecommunications Engineering 2013

quality especially for real-time applications. In [4], we have investigated how the two networks can co-operate, especially in terms of QoS, in order to reduce end-to-end delays and packet losses due to expiration. In this work, we extend [4] towards improving a part of the proposed mechanism which shares capacity to real time connections of the WiMAX network based on the use of Quality of Experience (QoE) metrics, to consider the overall performance of the system from the users' perspective. QoE expresses user satisfaction both subjectively and objectively, which are the two main categories of QoE metrics.

The paper is organized as follows. Section 2 presents an overview of using specific metrics for QoE management. Section 3 describes the basic architecture of our mechanism and the proposed improvement in the sharing of capacity to real-time traffic based on QoE metric, while section 4 contains the description of the simulation model used for evaluation purposes together with the obtained results. Finally, section 5 concludes the paper.

2   QoE Metrics for QoE Management

There is a large number of papers that use QoE metrics for measurement of video quality but very few that use these metrics for QoE management. [6] is the only work in UMTS that investigates the possibility of using QoE as a metric for scheduling decision. In order to get QoE feedback in real time, Pseudo-Subjective Quality Assessment (PSQA) technique is used [5], which is a hybrid approach between subjective and objective evaluation. PSQA metric starts by selecting the factors that may have an impact on the quality, such as: codec, bandwidth, loss, delay, and jitter. Then these factors are used to generate several distorted video samples, which are subjectively evaluated by a panel of observers. The results of the observations are then used to train a Random neural network (RNN) in order to capture the relation between the factors that cause the distortion (objective approach) and the perceived quality by real-human (subjective approach). In [6] loss rate (LR) of video packet and mean loss burst size (MLBS) are considered as the quality-affecting parameters for the training of RNN. MLBS parameter is the average length of a sequence of consecutive lost packets in a period of time and captures the way losses are distributed in the flow as this affects dramatically the perceptual quality of the video. After the training of RNN, Mean Opinion Score (MOS) is estimated in real time, which is the basic subjective metric, so that the scheduler can get MOS scores for making scheduling decision. [7] proposes a novel rate-adaptation mechanism based on QoE, using PSQA tool for obtaining MOS in real-time. The parameters used in PSQA are the loss rate of the I frames, loss rate of the P frames, loss rate of the B frames, and the MLBS of the I frames. The idea of the proposed scheme is to use QoE feedback from mobile stations to provision the current condition of the network and then adapt the rate accordingly. In [8], a novel packet scheduling algorithm for multi-hop wireless networks that jointly optimizes the delivery of multiple video, audio, and data flows according to the QoE metrics is developed. A previously proposed model to determine user satisfaction is used, where quality is given in terms of the objective metric Peak signal to noise ratio (PSNR), while MOS is produced through a non-linear curve mapping PSNR to MOS. The proposed scheduler locates sets of packet combinations across all active flows of all users that pass the node that would satisfy a given buffer reduction. For each of these combinations, an estimation of the user satisfaction expressed in MOS decrease for each flow is calculated. The scheduler then drops the packets whose combination results in the smallest decrease in QoE satisfaction based on a proposed cost function.

The target of this paper is to improve a previously proposed mechanism, in order to make the scheduling of Real-time Polling Service (rtPS) connections based on the use of QoE metrics. rtPS is the service in WiMAX that supports data streams consisting of variable-sized data packets that are transmitted at fixed intervals, such as MPEG video. QoE metrics are usually used for the assessment of the transmission of video on different network conditions, and rarely used in scheduling solutions, while they have never been used till now for scheduling in satellite networks. Subjective metrics are the most accurate for QoE measurements, as they are evaluated by real human. Their main shortcoming is that they are time-consuming and high-cost in man power. Thus, they cannot be easily repeated several times nor used in real-time (being a part of an automatic process). As the need for the proposed improvement is to be part of an automatic procedure, subjective and hybrid QoE metrics are excluded. From the already proposed solutions in other kind of networks, the solutions proposed in [6] and [7] have the drawback of using the PSQA metric for scheduling and QoE management. On the other hand, the solution proposed in [8] has increased complexity, as it calculates the QoE produced by every possible packet dropping. Our proposal aims to be simpler in order to be used in satellite networks, which have the drawback of delays. For all these reasons, the Media Delivery Index based on Frame Classification (FC-MDI) metric is chosen to be used in the existing mechanism, as it is an objective metric that gives a different weight to the loss of I, P, B frames which is useful for the scheduling of different categories of frames. The FC-MDI metric is an extension of the MDI (Media Delivery Index) metric [9], which is an objective metric that contains two numbers separated by colon: the delay factor (DF) and the media loss rate (MLR). DF is a time value indicating how many milliseconds' the buffer must be able to contain to eliminate jitter, while MLR is the computed difference between the number of media packets received during an interval and the number of media packets expected during an interval. Nevertheless, in MLR some important information is lost, such as whether the IP packets lost are consecutive or inconsecutive. It does not consider the quality degradation that suffered some propagated loss from previous temporally related frames, so [10] proposes FC-MDI which takes frame classification into account to improve the performance of the MDI measurement. It distinguishes the packet loss based on the frame classification, and gives in each frame a different weight. In all types of frames, the I-frame plays the most important role, as the rest frames of the whole group of pictures (GOP) cannot decode normally if the I-frame is lost. Compared with B-frame, P-frame relies less on its previous I-frames and P-frames.

3　Proposed Scheduling Solution

In [4], an interconnection of a satellite and a WiMAX network is proposed, assuming that one or more of the Return Channel Satellite Terminals (RCSTs) are also WiMAX

Base Stations (BSs) serving a number of Subscriber Stations (SSs). This integrated scheduling provision mechanism consists of three main parts: PartA is an entity at the RCST/BS that makes the capacity requests following a prediction-based approach, PartB is an entity at the Network Control Center (NCC) that allocates resources and creates the Terminal Burst Time Plan (TBTP), while PartC is an entity at the RCST/BS that shares the given capacity among its WiMAX subscribers. PartB accepts the capacity requests made from all PartAs, processes them and creates the TBTP in order to allocate the capacity of a superframe among the different RCSTs. PartC, located at the RCST/BS, contains the scheduling algorithm that is responsible to share the allocated capacity, to the uplink traffic arriving from the WiMAX network. In more detail, PartC classifies uplink traffic arriving from the SSs into five queues (UGS_queue, rtPS_queue, ertPS_queue, nrtPS_queue, BE_queue based on each packet's QoS service type). It then interprets TBTP (knows exactly which slots has been assigned to it) and selects which packets will be transmitted. This selection is made based on a priority scheme: it first selects packets from the UGS_queue, then from the rtPS_queue, then from the ertPS_queue, then from the nrtPS_queue and finally from BE_queue. Finally, it is also responsible to discard packets that are expired based on the deadlines set for their transmission to the satellite network and keep statistics on the packets transmitted and discarded. The *RTFS* (Real Time FIFO Scheduler) algorithm treats the transmission of packets of video connections with the logic of a First In First Out (FIFO) queue. The packets of all video connections are inserted in the rtPS_queue based on the order of their arrival. During the superframe, the PartC transmits, whenever it has available capacity based on the TBTP, the packets from this queue. A packet is dropped, if it has been expired due to delay. The performance of the mechanism was demonstrated in paper [4].

The target of this paper is to improve the *RTFS* algorithm, in order to make the scheduling of rtPS connections based on the use of FC_MDI QoE metric. The proposed *FC_MDI_S* algorithm makes the following procedures in the beginning of every superframe :

a) Dropping of the packets that are expired due to delay factor.
b) Computing of the *FC-MLR* value of every connection based on the loss of I, P, B frames in the previous superframe. The *FC-MLR* value is computed as follows :

$$FC - MLR = \frac{a * I_{PLoss} + \beta * P_{PLoss} + \gamma * B_{PLoss}}{interval},$$

where α, β, γ are weights with $(3 \geq \alpha > \beta > \gamma \geq 0, \alpha + \beta + \gamma = 3)$ and I_{PLoss}, P_{PLoss}, and B_{PLoss} are respectively the number of lost I, P and B frames.

c) Sorting of the video connections based on the *FC-MLR* value of the connections computed on step b under two versions. The first version is named *FC_MDI_SG* and has a greedy logic. In order to preserve the connections that have good quality, the connections are sorted based on *FC-MLR* value in ascending way, from the best quality to the worst. This will lead to the maintenance of the quality of some connections and the starvation of some other connections. The second version is named *FC_MDI_SF* and has a fair logic. In order to be fair and maintain all connections (even in worse quality), the connections are sorted in the opposite way than the previous version from the worst quality to the best.

$P_{3,3}$	$P_{1,3}$	$B_{2,1}$	$B_{3,1}$	$P_{3,2}$	$P_{2,2}$	$B_{1,1}$	$P_{1,2}$	$P_{2,1}$	$P_{3,1}$	$I_{3,1}$	$I_{2,1}$	$P_{1,1}$	$I_{1,1}$

Fig. 1a. Packets of three connections to the rtPS_queue of PartC

$B_{3,1}$	$B_{2,1}$	$B_{1,1}$	$P_{3,3}$	$P_{3,1}$	$P_{2,3}$	$P_{2,2}$	$P_{2,1}$	$P_{1,3}$	$P_{1,2}$	$P_{1,1}$	$I_{3,1}$	$I_{2,1}$	$I_{1,1}$

Fig. 1b. Transmission of packets under FC_MDI_SG algorithm

$B_{3,1}$	$B_{2,1}$	$B_{1,1}$	$P_{3,3}$	$P_{1,3}$	$P_{3,2}$	$P_{2,2}$	$P_{1,2}$	$P_{3,1}$	$P_{2,1}$	$P_{1,1}$	$I_{3,1}$	$I_{2,1}$	$I_{1,1}$

Fig. 1c. Transmission of packets under FC_MDI_SF algorithm

During the superframe, the PartC transmits whenever it has available capacity based on the TBTP. The *FC_MDI_SG* version transmits all the packets of one category, giving priority to I frames, then to P frames and last to B frames, and then moves on to packets of the same category of another connection. The order of the connections is the one computed to step c of the algorithm. On the contrary, the *FC_MDI_SF* version transmits one packet of one category from all connections, and then another packet of the same category from all connections, until exhausting all the packets of this category. After the transmission of all packets of the previous category, it moves on to the next category giving priority to I frames, then to P frames and last to B frames. The order of the connections is the one computed to step c of the algorithm. Figure 1 presents an example of transmission of packets under the previously proposed versions. In more detail, Figure 1a presents the frames of three connections as they have arrived in the rtPS_queue of PartC, Figure 1b presents the transmission of frames under *FC_MDI_SG* version and Figure 1c presents the transmission of frames under *FC_MDI_SF* version. The pointer i in the *Frame_Categor$_{yi,j}$* of Figure 1 shows the connection, while the pointer j shows the order of the packet of the specific frame category of i connection. Finally, the transmission of these packets as well as the dropping of the packets described in step a, are admeasured to the computing of the *FC-MLR* value of the connections for the next superframe.

4   Simulations

In order to measure the performance of the proposed algorithm, we accommodated the simulation program presented in [4]. The program is constructed in C++ and simulates the full operation of WiMAX network, as well as the DVB-RCS for the return link of a satellite network. We use the simulation scenario presented in [4] with three DVB-RCS terminals each one interconnecting a WiMAX network, all with the same number of subscribers. In the previous simulation scenario, every SS had multiple types of traffic, including video, compressed and uncompressed voice, ftp

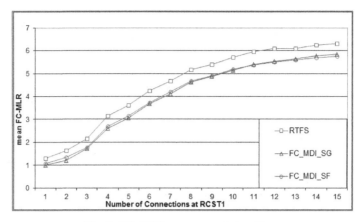

Fig. 2. FC_MLR per proposed algorithm

and http. In order to present the difference of the proposed mechanism regarding the QoE of the video connections, in the present simulation scenario every SS has only one video connection. The same video trace is used for every SS, in order to present the difference between the greedy and fair version. The source of this video trace is the "Alladin" film from "http://trace.eas.asu.edu/TRACE/ltvt.html" in high quality ("Verbose_Alladin.dat" file).

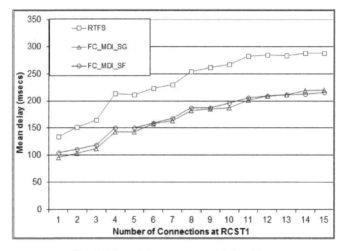

Fig. 3. Mean delay per proposed algorithm

The time frame length in WiMAX is set to 1msec, the packet size to 54 bytes and the modulation to 64-QAM for all SSs, leading to a transmission speed of 120Mbps (as indicated in the standard). The latency used in the WiMAX network for rtPS connections is 50msec.

The maximum transmission rate in the return link of the satellite network is 6Mbit/s, while the duration of the frame is set to 50msec and the superframe to

500msec, equal to the round trip delay. During the logon phase, each RCST terminal sets the CRA_level equal to zero (in order to present the difference between the quantity of the requested slots), the RBDC_max to 700kbps, the RBDC_timeout equal to 2 and the VBDC_max equal to 11 slots per frame. The latency used in the satellite network is 300msec.

Figure 2 presents the mean *FC-MLR* value for all connections of a SS of the two different algorithms. In the RTFS algorithm, the transmission as well as the dropping of packets are admeasured to the computing of the *FC-MLR* value of the connections for reasons of comparison. The proposed *FC_MDI_S* algorithm has lower values for *FC-MLR* than the *RTFS* one, which means that video connections under the proposed algorithm have better QoE.

Moreover, Figure 3 presents the mean delay of the proposed algorithm, which shows that both versions of the *FC_MDI_S* algorithm reduce considerably the mean delay of the connections. This was presumable, as the *RTFS* algorithm serves first the packets with the larger delay in the system, while the *FC_MDI_S* algorithm serves packets based on their frame category. The reduction in the mean delay is a substantially improvement, as we prefer video connections to have reduced delay.

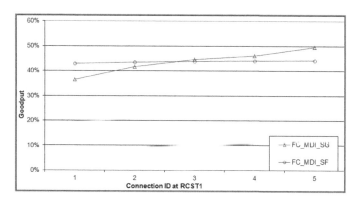

Fig. 4. Goodput per connection id for five connections per SS

The greedy and fair version of the proposed algorithm have the same performance concerning the goodput and mean delay, as the logic of the versions for sharing capacity is the same. They differentiate in the way they deal with the different connection identifiers. Figure 4 presents the goodput per connection identifier for five connections per SS. This figure shows the different performance of the greedy and the fair version, where the connections have differentiated goodpout under the greedy version and equal goodput under the fair version. In the fair version, the goodput is less than the goodput of the best connection and better than the goodput of the worst connection of the greedy version. It is due to the operator of the system to choose between them.

From the presented results, we conclude that the *FC_MDI_S* algorithm improves the QoE performance relatively to the *RTFS* algorithm, and it substantially improves the mean delay of the connections.

5   Conclusion

In this paper, we improve a previously proposed scheduling algorithm named *RTFS*. This algorithm is responsible to share the allocated capacity to the uplink traffic arriving from the WiMAX network in an integrated satellite/WiMAX network. After a bibliographic search for QoE metrics in WiMAX and satellite networks, the FC_MDI QoE metric is selected to be used in the proposed algorithm named *FC_MDI_S*. This is considered novel, as QoE metrics are mainly used for the assessment of video quality and not for scheduling. Especially in satellite networks, QoE metrics have never been used in management tools. We proposed and evaluated two versions for *FC_MDI_S*, and simulation results show that it considerably improves the QoE of video connections and reduces their mean delay.

References

1. IEEE Std 802.16-2004, IEEE Standard for Local and Metropolitan Area Networks – Part 16: Air Interface for Fixed Broadband Access Systems (October 2004)
2. IEEE Std 802.16e-2005, Amendment to IEEE Standard for Local and Metropolitan Area Networks - Part 16: Air Interface for Fixed Broadband Wireless Access Systems- Physical and Medium Access Control Layers for Combined Fixed and Mobile Operation in Licensed Bands (February 2006)
3. ETSI EN 301 790 V1.5.1, Interaction channel for Satellite Distribution Systems; DVB Document A054 Rev. 4.1 (January 2009)
4. Lygizou, A., Xergias, S., Passas, N., Merakos, L.: A prediction-based scheduling mechanism for interconnection between WiMAX and satellite networks. International Journal of Autonomous and Adaptive Communications Systems 2(2), 107–127 (2009)
5. Mohamed, S., Rubino, G.: A Study of Real–time Packet Video Quality Using Random Neural Networks. IEEE Transactions On Circuits and Systems for Video Technology 12(12), 1071–1083 (2002)
6. Piamrat, K., Singh, K.D., Ksentini, A., Viho, C., Bonnin, J.-M.: QoE-aware scheduling for video-streaming in High Speed Downlink Packet Access. In: Wireless Communications and Networking Conference (WCNC), April 18-21, pp. 1–6. IEEE (2010)
7. Piamrat, K., Ksentini, A., Bonnin, J.-M., Viho, C.: Rate Adaptation Mechanism for Multimedia Multicasting in Wireless Networks. In: Sixth International Conference on Broadband Communications, Networks, and Systems, BROADNETS 2009, September 14-16, pp. 1–7 (2009)
8. Reis, A.B., Chakareski, J., Kassler, A., Sargento, S.: Quality of experience optimized scheduling in multi-service wireless mesh networks. In: 17th IEEE International Conference on Image Processing (ICIP), September 26-29, pp. 3233–3236 (2010)
9. Krejci, J.: MDI measurement in the IPTV. In: 15th International Conference on Systems, Signals and Image Processing, IWSSIP 2008, June 25-28, pp. 49–52 (2008)
10. Fan, S., He, L.: A Refined MDI Approach Based on Frame Classification for IPTV Video Quality Evaluation. In: Second International Workshop on Education Technology and Computer Science (ETCS), March 6-7, pp. 193–197 (2010)

An Adaptive Connection Admission Control Algorithm for UMTS Based Satellite System with Variable Capacity Supporting Multimedia Services

Anju Pillai, Yim Fun Hu, and Rosemary Halliwell

University of Bradford, United Kingdom
{a.pillai,y.f.hu,r.a.halliwell}@Bradford.ac.uk

Abstract. This paper is focused on the design of an adaptive Connection Admission Control (CAC) algorithm for a Universal Mobile Telecommunication System (UMTS) based satellite system with variable link capacity. The main feature of the proposed algorithm is to maximize the resource utilization by adapting to the link conditions and the antenna gain of the users. The link quality of the user may vary depending on the weather condition, user mobility and any other propagation factors. The algorithm is compared against a non-adaptive admission control algorithm under different test cases. The proposed CAC algorithm is simulated using MATLAB and the performance results are obtained for a mix of multimedia traffic classes such as video streaming, web browsing, netted voice and email. The simulation results indicate a higher system performance in terms of the blocking ratio and the number of admitted connections.

Keywords: Connection Admission Control, MATLAB, UMTS, multimedia traffic.

1   Introduction

The Connection Admission Control (CAC) is an integral resource management scheme for providing Quality of Service (QoS) in a network. Satellite networks have been growing in popularity owing to their large geographic coverage, and fast deployment when compared to terrestrial networks. Although a satellite system can provide added advantages to the telecommunication infrastructure, the transmission capacity of a satellite is very limited as compared to that of terrestrial networks. Hence, the radio resources have to be managed efficiently such that an acceptable quality of service is delivered. The design of an efficient CAC scheme for satellite networks has been extensively dealt in the literature. A Double Movable Boundary Strategy (DMBS) resource allocation scheme is proposed in [1]. It dynamically controls the boundary policy of the resource sharing amongst the different traffic class according to the variable network load conditions. CAC and the bandwidth allocation decisions are taken at the beginning of each control period. The impact of the queue threshold value on the performance of the DBMS allocation policy is evaluated.

P. Pillai, R. Shorey, and E. Ferro (Eds.): PSATS 2012, LNICST 52, pp. 9–16, 2013.

A predictive CAC algorithm is proposed in [2] for onboard packet switching satellite systems. The algorithm performs the online measurements for the established connections. Based on the estimated parameters, the individual cell loss ratio (ICLR) is predicted ahead of the current time which is used in the CAC decision process. A measurement based admission control (MBAC) for onboard processing satellite is also proposed by the authors of [3]. Unlike, [2] where CAC is implemented onboard the satellite, this paper proposes a scalable CAC implemented on ground in Network Control Centre (NCC). The on-board measurements are performed on aggregate traffic rather than on single traffic for each downlink to reduce the computation complexities on the satellite, which are transmitted to the NCC. The CAC is based on the computation of the downlink effective bandwidth which is an estimation of the actual downlink bit rates used by the in-progress connections. Unfortunately, these approaches are only suitable for a system where the link capacity remains fixed. However, for the system in consideration, the link capacity varies with the antenna gain and the changing link condition of the user which in turn may vary depending on the weather conditions, user mobility and any other propagation factors. This paper proposes an adaptive CAC algorithm for a UMTS based satellite system which adapts to the varying link capacity in order to maximize the resource utilization. The adaptive CAC algorithm is compared against a non-adaptive CAC algorithm which does not consider these factors into account while making the CAC decision.

The paper first describes the network architecture of the system followed by the presentation of the CAC functional model for the admission control algorithm. Section 4 describes the proposed adaptive CAC algorithm and the non-adaptive CAC algorithm used for comparison. The simulation results, generated in MATLAB, are then presented and analyzed. Finally, conclusions and further work are discussed.

2   Network Architecture

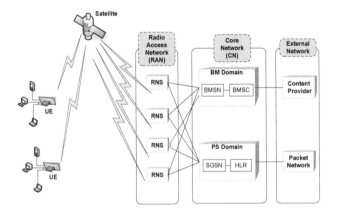

Fig. 1. UMTS based satellite network architecture

The network architecture of a UMTS based satellite system is shown in Fig. 1. It is divided into three segments: a) The User Equipment (UE) segment consists of a portable satellite modem, the Mobile Terminal (MT), connecting to a Terminal Equipment (TE) such as a personal computer or a PDA, allowing users access to UMTS services. Multiple TE can be connected to one MT such that multiple data connections can belong to one MT; b) The ground segment consists of the Radio Access Network (RAN) and the Core Network (CN). The CAC controller is located in the Radio Network Controller (RNC) of the RAN; c) The satellite segment consists of a multi-beam geostationary satellite system that provides a transparent link between the UE and the RNC. MF-TDM and MF-TDMA are adopted in the forward (satellite-to-user link) and the reverse (user-to-satellite link) links respectively. In the forward direction, each satellite channel has a bandwidth of 200 kHz, which is termed as forward sub-bands.

The proposed adaptive CAC algorithm focuses on the resource availability in the forward direction using a fixed number of forward sub-bands to admit the data connections. Each MT is tuned to a particular forward sub-band and therefore, all the data connections belonging to a MT are also tuned to the same forward sub-band. The system supports different MT Classes pertaining to the size of their antennas and operating scenario of the MT such as portable, land-vehicular, maritime or aeronautical.

3   CAC Functional Block Model

Fig. 2. shows the functional model of the proposed adaptive CAC algorithm represented by the CAC Processor. The CAC Processor is an event driven functional entity which encompasses the admission control algorithms. The algorithm is executed at the arrival of a new connection request and decides whether the connection request can be admitted. The difference between the two algorithm lies in the way the resource consumed by a connection is calculated.

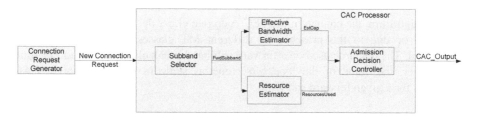

Fig. 2. Functional block model of the CAC algorithms

The Connection Request Generator generates different types of connection requests and sends the request to the CAC Processor. The model consists of four functional blocks: a Subband Selector, an Effective Bandwidth Estimator, a Resource Estimator and an Admission Decision Controller. The Subband Selector selects the forward sub-band for a new MT from the list of available sub-bands. Two methods have been proposed: a) *MinConnSubSel* selects the forward sub-band with the minimum number of connections running from the list of available sub-bands. This method allows a basic form of load balancing, b) Random method randomly selects a forward sub-band. The Effective Bandwidth Estimator estimates the bandwidth required by the connection, *EstCap*, based on the traffic class, priority, source utilization and QoS value of the connection. The Resource Estimator calculates the total resources consumed, *ResourcesUsed*, by the active connections on the given forward subband. The admission decision controller produces an output, *CAC_Output*, which may indicate an admission or a rejection of a connection request.

4   Admission Control Algorithms

The difference between the two admission control algorithms; the adaptive and the non-adaptive, lies in the calculation of the resources consumed by a connection as explained below.

4.1   Adaptive

The total resource consumption on a given forward subband is calculated using the FEC code rate applied in the physical frame for a given MT. The FEC code rate varies with the changing link condition and the class of the MT. Each MT class supports a range of FEC code rates. The adaptive CAC checks the code rate against the class of the MT for the given link condition. The resource used on a forward sub-band is calculated as follows:

$$ResourcesUsed_{givenfwdsubband} = \sum_{allconnections} EstCap \, / \, coderate$$

(1)

where, the '*coderate*' value varies constantly in adaptive CAC algorithm.

4.2   Non-adaptive

For this algorithm, the change in the FEC code rate either due to a change in the link condition or due to the presence of different MT classes in the system is not considered. The resource used on a forward sub-band is calculated similar to Equation 1; however, the value of the '*coderate*' remains static in the non-adaptive CAC algorithm for a given forward sub-band.

5   Simulation Results and Analysis

The performance of the adaptive and the non-adaptive admission control algorithms are analyzed and compared using a mixed class of multimedia traffic. The performances are measured under different test cases and the results are compared Table 1 summarizes the MATLAB scenario configuration set for the simulation.

Table 1. Simulation parameters for scenarios

Common Simulation Parameters		Values
Video Streaming	Number of connections	25
	QoS (kbps)	32
	Source Utilization	0.8
	Mean Burst Period	0.1
	Avg. Holding time (sec)	300
Netted Voice	Number of connections	25
	QoS (kbps)	60
	Source Utilization	0.6
	Mean Burst Period	0.01
	Avg. Holding time (sec)	240
Web Browsing	Number of connections	25
	QoS (kbps)	32
	Source Utilization	0.4
	Mean Burst Period	0.01
	Avg. Holding time (sec)	200
Email	Number of connections	25
	QoS (kbps)	120
	Source Utilization	0.2
	Mean Burst Period	0.01
	Avg. Holding time (sec)	150

5.1   Test Case 1: Effect of Link Condition

Fig. 3 and Fig. 4, show the effect of link quality on the blocking ratio and the number of admitted connections respectively. For the adaptive algorithm, the blocking ratio reduces under high link quality condition as compared to low link quality condition. The code rates supported in the high link condition are higher than in the low link condition and since the higher code rates allow more data to be sent in the physical frame, more number of connections can be admitted in turn reducing the blocking ratio. For the non adaptive algorithm, a fixed code rate pertaining to the lowest code rate supported, by the given type of forward subband, is considered. This results in a reduced amount of data that can be sent in a frame in the physical layer and hence reduces the number of connections that can be admitted, which in turn increases the blocking ratio.

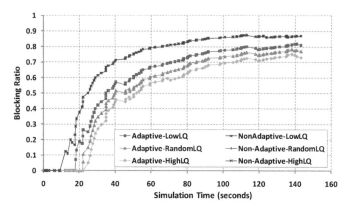

Fig. 3. Effect of link condition on the blocking ratio

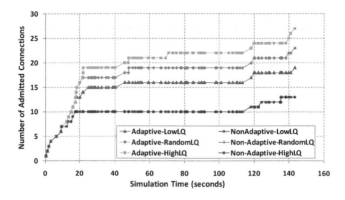

Fig. 4. Effect of link condition on the number of admitted connections

5.2   Test Case2: Effect of Antenna Gain

Fig. 5 and Fig. 6, show the effect of the antenna gain on the blocking ratio and the number of admitted connections respectively. For the adaptive algorithm, the blocking ratio varies with the different MT classes. For this test case, all the MTs are assumed to be in the low link condition. Each MT class supports a range of code rates. The MTs belonging to class 1 supports a higher code rate than class 2 followed by class 3, for the given low link condition. A higher code rate allows more data to be sent in the physical frame leading to an increase in the number of connections admitted and in turn reducing the blocking ratio. For the non-adaptive algorithm, the blocking ratio and the number of admitted connections remains same irrespective of the type of MT classes used as the algorithm considers a fixed code rate for a given forward subband.

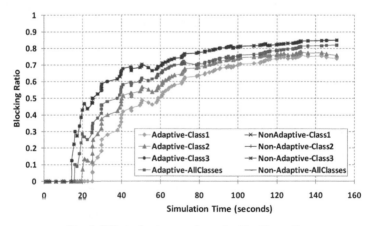

Fig. 5. Effect of antenna gain on the blocking ratio

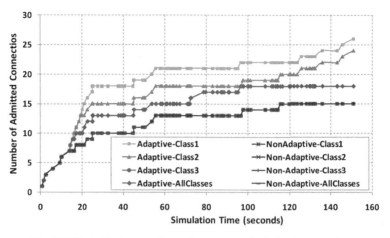

Fig. 6. Effect of antenna gain on the number of admitted connections

5.3   Test Case 3: Effect of Traffic Intensity

Fig. 7 and Fig. 8, show the effect of the traffic intensity on the blocking ratio and the number of admitted connections respectively. For this test case, all the MTs are assumed to be in the low link condition. The traffic intensity or the offered traffic load (ρ) is defined as the ratio of the average arrival rate ('λ') to the average service rate ('μ'). Therefore, $\rho=\lambda/\mu$. The system runs with different traffic load by changing the average inter arrival times for the video streaming traffic. As can be seen, the blocking ratio for the adaptive algorithm remains lower than the non-adaptive algorithm as the traffic load increases. The amount of resources needed to accommodate the same amount of traffic for the non-adaptive algorithm is much higher than that for the adaptive algorithm with the increase in the traffic load and thus leads to a higher blocking ratio.

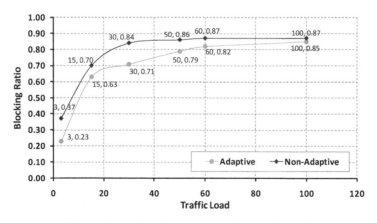

Fig. 7. Effect of traffic intensity on the blocking ratio

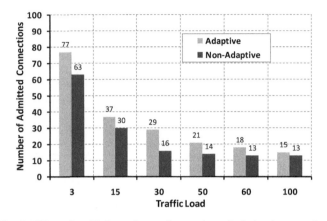

Fig. 8. Effect of traffic intensity on the number of admitted connections

6   Conclusion and Future Work

In this paper, an adaptive admission control algorithm has been presented for a UMTS compatible satellite system. The proposed algorithm allows the system to maximize the resource utilization by adapting to the variable link capacity of the system caused by the changing link condition of the users and the different antenna gains supported by the system. The algorithm is compared against a non-adaptive admission control algorithm under different test cases. The simulation results indicate a higher system performance in terms of the blocking ratio and the number of admitted connections. This work will be extended in the future to support the multicast traffic. It will involve defining an admission decision process for the multicast traffic and utilizing the adaptive admission control algorithm and the functional model defined in this paper to include the support of the multicast traffic and thereby testing the system under different test cases.

References

1. Koraitim, H., Tohme, S.: Resource Allocation and Connection Admission Control in Satellite Networks. Selected Areas in Communications, IEEE Journal 17, 360–372 (1999) ISSN: 0733-8716
2. Jang, Y.M.: Estimation and Prediction-Based Connection Admission Control in Broadband Satellite systems. ETRI Journal 22, 40–50 (2000)
3. Priscoli, F.D., Pietrabissa, A.: Control-based Connection Admission Control and Downlink Congestion Control procedures for satellite networks. Journal of Franklin Institute 346(9), 923–944 (2009)
4. 3GPP TS 23.107: Quality of Service (QoS) concept and architecture(2006)
5. Ahmed, M.H.: Call Admission Control in wireless networks: A comprehensive survey. IEEE Communications Survey and Tutorials 7, 1553–1877 (2005) ISSN: 1553-877X

Network Coding for Next Generation Personal Satellite Converged Services

Daniel E. Lucani[1,*] and Marie-José Montpetit[2]

[1] Instituto de Telecommunicações, DEEC, Faculdade de Engenharia da
Universidade do Porto, Porto 4200-465, Portugal
[2] Research Laboratory of Electronics, Massachusetts Institute of Technology,
Cambridge, MA, USA
`dlucani@fe.up.pt, mariejo@mit.edu`

Abstract. Seeking to meet the resilience, efficiency, and quality of experience challenges of personal and converged satellite services, we present a new approach that leverages the benefits of network coding. The salient features of this strategy are (i) source nodes manage and control the transmission of linear combinations of data packets through heterogeneous communication routes, and (ii) intermediate nodes at each route can generate new coded packets with opportunistic storage. Hence, the amount of data and redundancy sent through each route can meet the required performance of the different sessions. In particular it can help surmount varying channel conditions and correct erasures but also adapt to the different delays and bandwidth that are features of the converged PSAT networks of the future. The main technical challenge is to choose adequate, coding-aware policies to leverage heterogeneous networks based on content and user requirements. We present preliminary analysis that illustrates that exploiting the routes jointly can be performed seamlessly using network coding even with limited feedback capabilities.

Keywords: Convergence, Heterogeneous Networks, Network Coding, Satellite Communications.

1 Introduction

Following well-established trends in the terrestrial networks, personal satellite services (PSATS) will provide end users with a variety of personalized services focused on rich media content. While there is still a difference in terms of spectrum between fixed and mobile satellite systems, the same convergence that was experienced in terrestrial networks in terms of services is happening in the satellite realm. The aim there is to provide services that are independent of the satellite system that is serving them and leverage terrestrial networks when appropriate. Advances in terrestrial and space segments' technology already allow personalized multimedia and broadband services to be offered from a single satellite system, such as the Inmarsat Global Xpress [1], and there are technical and

---

* This work was supported in part by the Instituto de Telecommunicações COHERENT
Project, under the FCT project PEst-OE/EEI/LA0008/2011.

P. Pillai, R. Shorey, and E. Ferro (Eds.): PSATS 2012, LNICST 52, pp. 17–25, 2013.

commercial incentives to continue integrating satellite and terrestrial networks, as demonstrated by systems like TerreStar [2]. Recent advances in networking technologies offer novel approaches for this seamless platform integration. In this paper, network coding will be presented as a key enabler for i) improving the performance of IP-based protocols over satellite, but also for ii) empowering network combining itself be it satellite-satellite of the fixed or mobile category or satellite-terrestrial when appropriate. We infer that the resulting services strategies are instrumental in securing the position of satellite networks in the next generation Internet.

Network coding (NC) considers digital traffic as algebraic entities not just data that needs to be transported [3]. Because these entities can be multiplied by constants and linearly combined, some of the usual networking constraints such as state information and independent use of network resources can be lifted. In addition since the coding parameters can be simple random numbers over a fairly short Galois field [4] the coding and decoding process may remain simple [5]. While network coding is not specific to a single type of network, it has shown promise on long delay networks such as satellite and underwater networks [6].

In the remainder of the paper we present our approach to satellite network convergence based on NC. We will show that it allows soft service-focused handover and heterogeneous network combining with overall improved performance. Section 2 reviews some aspects of satellite system convergence leading to our proposed architectures, outlined in Section 3. Section 4 provides a very short introduction to network coding that clarifies aspects of the analysis presented in Section 5. This analysis is based on heterogeneous path analysis and probabilistic principles. Section 6 concludes with some look into the future especially referring to the position of satellite systems for content dissemination in the Future Internet.

2   Convergence in Satellite Networks

Satellite networks are impacted by the disruptions in the communications industry, driven by an emphasis on Internet based services, machine-to-machine communications, and ubiquitous multimedia. In this environment, the converged nature of many services is driving innovation including systems that combine satellite and terrestrial 3G/4G to guarantee Internet access everywhere. The successful integration of satellite networks into the Internet has been progressing for over 15 years and has lead to implementable standards such as those defined in the IETF TCPSAT [7] and IPDVB [8] as well as DVB-RCS [9] and the ETSI-BSM [10].

Moving further, converged networking architectures are characterized by the unbundling of services and underlying transport functions (i.e. network technologies) which enables the definition of service-driven network policies independently of the underlying infrastructure. One architectural embodiment of the principle is provided by the Next-Generation Networks (NGN) architecture, one that has already been adapted to broadband services via fixed satellite systems [11] [12]. While these are major steps towards satellite-terrestrial hybrid networks, there remains open challenges to respond in terms of capacity, packet erasures, and delay mismatches between systems. Full convergence of the satellite network themselves has not being fully investigated but is an active field of research. There are many services, from infrastructure support to social television and emergency management, which require revisiting the way satellites and other networks can work together. The new challenge is to develop mechanisms to exploit collaboration and interaction across the different domains beyond exchanges of signaling information.

This is where NC can clearly make a difference. In order to profit to the maximum of the use of NC we need to investigate the architecture of the network that will support it from the source nodes that implement the content mixing to the judicious placement of intermediate nodes, which without the need of decoding, can remix different flows and guarantee they will be delivered together to the receiver. The decoding can be naturally pushed all the way to the end user device. This architecture is presented in the next section.

3   Architecture and Derived Scenarios

A hybrid architecture that combines both space and ground segments is considered in this paper. Figure 1 presents an overview of the targeted multi-system and converged ecosystem. The space segment can be one or more fixed and mobile systems and the ground segments, while in principle of any type, will be assumed to be mostly wireless and broadband in nature. Specific scenarios can be carved from this generic architecture that match service characteristics or operational requirements. In all our scenarios, we will however assume that the NC elements remain in the terrestrial domain. While the operations needed to perform network coding are linear in nature and require little processing, their implementation in a regenerative satellite are beyond the scope of this paper

Hybrid architectures are however plagued by the consequences of heterogeneity. For example, different paths to the same destination may experience differing delays, fading and erasures, not all network capacity comes at the same operational cost, etc. NC will reduce the complexity of managing and operating this hybrid network by enabling mixes of data to be sent through different paths and be recombined at the end without having to keep packet trackers and otherwise state information. As will be seen in the next section, even without implementing a feedback loop to manage the mixing of packets, significant gains can be made when compared with non-network coded system. With NC the capabilities of each system, such as capacity or cost of the bandwidth, can be traded, combined and optimized. In our architecture, the source nodes (the data sources) implement the initial network coding and the information can be sent via a combination of terrestrial and satellite paths who will act as partners in the delivery of the information. Intermediate nodes, like satellite ground stations, receivers and wireless nodes can act as decoders or as re-encoders based on the instantaneous conditions they experience but also delay requirements or user equipment capabilities.

Use cases can be defined based on the elements of Figure 1. The first one is system combining. With network coding, a satellite system can complement another one in times of fading. These could be of the fixed-fixed, fixed-mobile or mobile-mobile category depending on equipment, business models and regulatory environments. Even if the second satellite is of lower capacity, the combination of the two systems can allow a session/terminal to survive a momentary fading and loss of capacity without significant QoS degradation and facilitate a handover if appropriate. As will be seen in the next section there is a high probability that the rich mix of content coming from a secondary path will contain enough mixed packets (degrees of freedom) to recuperate lost or delayed information on the primary path. There is evidence coming from terrestrial networks that only a few packets borrowed from a secondary path can greatly improve overall performance [13].

Another use case could combine a wireless/mobile terrestrial network while leveraging a fixed satellite network; this can add interactivity and support rich video applications. It will be seen in the analysis (Section 5) that with network coding this can be implemented with little complexity. Since, network coding relaxes the need for maintaining session state and data pointers in relaying nodes, it supports this combination of fixed satellite and terrestrial networks. As can be foreseen from these two very simple use cases, NC can provide additional the coverage to increase the number of satisfied customers.

4   Network Coding Primer

Network coding (NC) offers exciting possibilities for the efficient transmission of video over wireless and bottleneck networks by considering traffic as algebraic information and sending linear combinations of packets. Overall NC allows to reduce the required number of transmissions to complete a file or stream operation over noisy or unreliable networks. This results in increased throughput but also, since the smaller completion time will reduce application layer timeouts, of goodput which is a strong measure of Quality of Experience (QoE). While NC adds some complexity to both source and destination nodes since it involves performing linear operations these are not too demanding and have been successfully implemented in hand-held devices [5]. Also the use of codes defined on small Galois fields (GF), and of codes that are systematic, i.e., packets are initially sent uncoded followed by a number of mixed packets, can significantly reduce coding operation complexity [14].

We illustrate the idea of network coding in a simple multiple route scenario in Figure 2 and compare it to a more traditional technique. A classical system trying to include additional redundancy could transmit by assigning resources in both routes and sending the same packets through each of the beams, as in Figure 2 (a). This decision is agnostic to the underlying channel conditions and is meant to provide additional reliability when a single packet transmission does not fulfill the required reliability. In a network coded system, we have the flexibility to send a different fraction of the data through each beam (as long as enough coded packets are sent) and to choose the desired level of redundancy. In Figure 2 (b) the system chooses to send 3 coded packets through one route and 2 through the other due to channel constraints and/or system load.

Although the system of Figure 2 (a) sends one packet more than its coded counterpart of Figure 2 (b), it is simple to see that the coded system provides

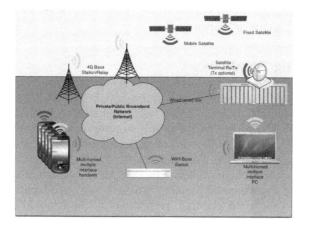

Fig. 1. Overview

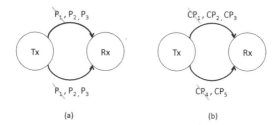

(a)   (b)

Fig. 2. A simple example of multiple routes. (a) Example of a transmission that sends the same 3 data packets through each route for increased reliability. (b) Example of a network coding approach where coded packets are sent through each route.

higher resiliency to packet losses. In our example, both cases may sustain up to two packets being lost. However, without coding when the same packet is lost in both routes, that packet is simply not recovered. Our coded example does not share this problem as different linear combinations can be sent through each route, guaranteeing resilience to exactly 2 packet transmissions, since the receiver only requires 3 independent linear combinations (out of 5 that were generated) to recover the original data.

From the example of Figure 2 (b) it is clear that a coded mechanism can provide additional guarantees of recovering all data packets. However, if more losses occur it may impede the receiver from recovering any single packet. In RLNC, coding across M packets requires M coded packets to recover any information. The key is to find a trade-off between partial recovery of the data by using a sparser code (RLNC is a dense code) and the inherent loss in performance due to the sparser nature of the code (RLNC is delay-optimal in our example for large enough field). A simple solution is to use a systematic structure, i.e., original packets are sent without coding once, while all additional packets are sent coded with RLNC [14]. Systematic network coding provides no degradation in performance while ensuring i) partial recovery of the packets, and ii) a reduction in decoding complexity, as shown in [14].

5   Motivating Analysis

Consider the problem of transmission of data packets from a source to a destination in a time-slotted system, where two independent channels are available. At each time slot, the source can transmit random linear network coded packets through both channels (different coded packet in each), one channel, or can decide to not transmit in that time slot.

We assume an independent Gilbert-Elliott model for the channels, where $p_{(g,i)}$ and $p_{(b,i)}$ are the corresponding packet erasure (loss) probabilities on the i-th link for the good and bad channel, respectively. The probability of link i to remain in state $c \in \{b, g\}$ is given by $p_c^{(i)}$.

We assume that a genie indicates the state $C = (c_1, c_2)$ of the two channels, i.e., the probabilities of packet loss in each channel, at each time slot. However, the event of a packet loss is not known *a priori* to the genie.

Let us model the general state of the system at the n-th time slot as $\mathcal{S}(n) = (Q(n), c_1(n), c_2(n), \mathcal{P}(n))$, where $Q(n)$ represents the number of independent linear combinations (or degrees of freedom) missing at the receiver at time slot n, $c_1(n)$ and $c_2(n)$ represent the state of the Gilbert-Elliott channels, and $\mathcal{P}(n)$ constitutes the source's policy. We assume an online NC approach as in [15].

We define a policy $\mathcal{P} \in \{\emptyset, S_1, S_2, \{S_1, S_2\}\}$ as a function that schedules the transmission of each packet based on the channel states of the available routes, where S_i represents the event of the source transmitting through channel i.

5.1   Optimal Policy in Terms of Reduction of Channel Utilization

In the following we define an optimal policy in terms of the reduction of the mean channel utilization, using a similar technique to [16]. Let us first define the *effective* mean erasure probability of S_i, $\bar{p}_{\text{eff}}(i)$ as the mean erasure probability seen by the transmitter as a result of the scheduling policy. We define $\alpha_{(i,C)}$ and $Pr_{(i,C)}$ as the fraction of the rate of S_i and the probability of transmission through channel i during the channel state C, respectively. Thus,

$$\bar{p}_{\text{eff}}(i) = \sum_{C=(c_1,c_2)\in\{g,b\}^2} p_{(c_1,i)} Pr_{(i,C)} \pi_C, \tag{1}$$

where π_C constitutes the stationary probability of the channel state C, which can be easily determined through standard finite Markov chain techniques. The channel utilization of channel i in our system is given by

$$U_i([Pr_{(i,C)}]) = \sum_C Pr_{(i,C)} \pi_C. \tag{2}$$

The optimization problem is stated as:

$$\min_{[Pr_{(i,C)}]} \sum_i U_i \left(Pr_{(i,C)} \right),$$

subject to

$$Pr_{(i,C)} \in [0,1], \quad \forall C \in \{g,b\}^2, i \in \{1,2\}$$

$$\sum_{i \in \{1,2\}, C \in \{g,b\}^2} \alpha_{(i,C)} = 1,$$

$$(1 - p_{(c_1,1)})Pr_{(1,C)}\pi_C = \lambda\alpha_{(1,C)}, \quad \forall C \in \{g,b\}^2,$$

$$(1 - p_{(c_2,2)})Pr_{(2,C)}\pi_C = \lambda\alpha_{(2,C)}, \quad \forall C \in \{g,b\}^2.$$

In the first line, we have used the right hand side of (1) as the argument of $U_i(\cdot)$. The last two conditions capture the fact that the probability of S_k transmitting in a given channel state is linked to the mean usage of the channel during that state, e.g., $\lambda\alpha_{(1,C)}/(1 - p_{(c_1,1)})$ for channel 1. We emphasize that to achieve this throughput performance using no NC would require a feedback mechanism that signals the correct reception of packets sent at each time slot. On the other hand, NC can achieve this performance naturally without requiring such an intensive feedback mechanism. In fact, feedback is required for other practical purposes, for example, i) to maintain reasonably sized queues at the sender while reducing decoding complexity [15], or ii) to enforce a required delay for decoding the packet at the receiver [17].

The optimal policy for a given channel state C and source rate λ is given by the vector $[Pr_{(i,C)}]$ that results of this optimization. The optimal policy $\mathcal{P}_{opt}(C) \in \{\emptyset, S_1, S_2, \{S_1, S_2\}\}$ can be stated as

Policy 1

$$\mathcal{P}_{opt}(C) = \begin{cases} \{S_1, S_2\} & w.p. \ Pr_{(1,C)}Pr_{(2,C)} \\ S_1 & w.p. \ Pr_{(1,C)}(1 - Pr_{(2,C)}) \\ S_2 & w.p. \ (1 - Pr_{(1,C)})Pr_{(2,C)} \\ \emptyset & otherwise \end{cases}$$

where S_i indicates the event of transmitting through channel i. Note that the probability of transmitting through channel 1 and channel 2 is independent, thus transmission over two channels or no channels at each time slot is possible.

This policy is probabilistic in nature and relies on the fact that the queues are assumed to be of infinite size. The latter allows the system to store the (coded) packets while awaiting a good channel. In practice, deterministic algorithms that link the queue lengths and channel-awareness may be more relevant since they may guarantee better delay performance, albeit with a possibly degraded throughput region.

6 Conclusion

This paper presents network coding as a key enabler for personal satellite converged services. It emphasizes an architecture for these satellite networks that maximizes the benefits of network coding by exploiting multiple available routes in the space and terrestrial segments and opportunistic coding at various locations in the network. The crux of the solution lies in choosing coding-aware policies that leverage (i) statistics from the available, time-varying channels, and (ii) content and user requirements to allocate the right level of redundancy at each transmission route as a means to control the performance of each session. A case study with Markovian channels is reported; it emphasizes an optimal probabilistic policy that performs transmission decisions based on a joint state of the routes available to the transmission. This result represents a very promising starting point for more in-depth evaluation of the potential benefits of network coding for converged satellite services. Future research will analyze the different available channels jointly as well as adding ancillary information to the policies, e.g, statistics of the channels, knowledge of traffic statistics, feedback etc, to provide practical mechanisms that can be seamlessly implemented in current and future systems. By enabling novel combination of systems and fixed-mobile as well as satellite-terrestrial convergence NC will allow operators to flexibly define new services and refine current offerings for optimized performance in terms of cost, delay and other quality of service parameters that are required by their customers.

Acknowledgements. The authors would like to acknowledge the ETSI Satellite Earth Station and Systems (SES) Working Group for initiating the discussion on fixed-mobile satellite networks and supporting work on satellite-terrestrial integration in the NGN, in particular R. Mort, R. Goodings, A. Noerpl, and N. Chuberre. We would also like to thank Marcus Vilaça and David Bath of Inmarsat, and Dave Bettinger of iDirect for their encouragement in pursuing this work. Finally we would like to thank our anonymous reviewers for their comments leading to this version of the paper.

References

1. Global Xpress, http://www.inmarsat.com/About/Newsroom/00037147.aspx?language=EN&textonly=False
2. TerreStar, http://www.terrestar.com/
3. Koetter, R., Médard, M.: An Algebraic Approach to Network Coding. IEEE/ACM Transactions on Networking 11(5) (2003)
4. Ho, T., Koetter, R., Médard, M., Karger, D.R., Effros, M.: The Benefits of Coding over Routing in a Randomized Setting. In: IEEE International Symposium on Information Theory (2003)
5. Heide, J., Pedersen, M.V., Fitzek, F.H.P., Larsen, T.: Network Coding for Mobile Devices - Systematic Binary Random Rateless Codes. In: IEEE International Conference on Communications (ICC) - Workshop on Cooperative Mobile Networks, Dresden, Germany (2009)
6. Lucani, D.E., Médard, M., Stojanovic, M.: On Coding for Delay - New Approaches Based on Network Coding in Networks with Large Latency. In: Proc. ITA Workshop, San Diego (2009)
7. IETF TCP over Satellite (TCPSAT), http://datatracker.ietf.org/wg/tcpsat/charter/

8. IETF IP over DVB - IPDVB Working Group, http://datatracker.ietf.org/wg/ipdvb/charter/
9. ETSI EN 301 790 v.1.4.1: Digital Video Broadcasting (DVB); Interaction channel for satellite distribution systems (2005-09)
10. ETSI TR 101 984 v1.1.1: Satellite Earth Stations and Systems (SES); Broadband satellite multimedia; Services and Architectures (2002-2011)
11. ETSI ES 282 001 v3.4.1: Telecommunications and Internet converged Services and Protocols for Advanced Networking (TISPAN); NGN Functional Architecture (2009)
12. ETSI TS 102 855 v.1.1.1: Satellite Earth Stations and Systems (SES); Broadband satellite multimedia; Interworking and Integration of BSM in Next Generation Networks (NGNs) (2011-03)
13. Kulkarni, A., Heindlmaier, M., Traskov, D., Medard, M., Montpetit, M.J.: Network Coding with Association Policies in Heterogeneous Networks. In: Proc. NC-Pro 2011 (May 2011)
14. Lucani, D.E., Médard, M., Stojanovic, M.: Systematic network coding for time-division duplexing. In: Proc. IEEE International Symposium on Information Theory, Austin, TX, USA, pp. 2403–2407 (2010)
15. Sundararajan, J.K., Shah, D., Médard, M.: ARQ for network coding. In: IEEE International Symposium on Information Theory, Toronto, Canada, pp. 1651–1655 (2008)
16. Lucani, D.E., Kliewer, J.: On the Delay and Energy Performance in Coded Two-Hop Line Networks with Bursty Erasures. In: Proc. IEEE International Symposium on Wireless Communication Systems, Russia (2011)
17. Lucani, D.E., Médard, M., Stojanovic, M.: Online Network Coding for Time-Division Duplexing. In: IEEE Global Telecommunications Conference, Miami, FL, USA, pp. 1–6 (2010)

Performance Analysis of an Enhanced Spread Spectrum Aloha System

Florian Collard[1,2], Annamaria Recchia[1], Natalia Antip[1],
Antonio Arcidiacono[1], Daniele Finocchiaro[1], and Orazio Pulvirenti[1]

[1] Eutelsat S.A., 70 rue Balard, 75502 Paris Cedex 15, France
{fcollard,arecchia,nantip,aarcidiacono,
dfinocchiaro,opulvirenti}@eutelsat.fr
[2] ISAE Campus Supaero, 10 Avenue Edouard Belin, 31055 Toulouse, France

Abstract. In this paper we present tests results of the first platform implementing a high performance Random Access protocol dubbed Enhanced Spread Spectrum Aloha (E-SSA), conceived for satellite applications. The aim of the study is to highlight key features of the system, introducing experimentally-found threshold values and translating them into operating conditions. A system overview is provided together with a summary of the most relevant parameters and the framework of the tests. The preamble detection and the packet demodulation processes are analysed, showing the importance of exploiting Successive Interference Cancellation techniques (SIC). It is shown that the code collision, due to a robust rate in the turbo-code and to the iterative cancellations, does not affect the system performances. Finally, Packet Loss Ratio (PLR) in presence of received carrier power and frequency unbalance is detailed with the aim of simulating more realistic scenarios.

Keywords: Random Access, Enhanced Spread Spectrum Aloha, Successive Interference Cancellation, Satellite Messaging Protocol.

1 Introduction

The launch of W2A in April 2009, carrying into orbit the first commercial S-band payload over Europe, represents a great opportunity for space telecommunications industry, given its innovative nature in the satellite domain. The MSS S-band frequencies, residing in the 2GHz band adjacent to 3G UMTS, allow the use of small omni-directional antennas and enable the deployment of satellite/terrestrial hybrid networks. Furthermore, the large reflector on-board requires a moderate EIRP at the terminal side, thus opening the way for a full exploitation of the return link (terminal-to-satellite) [1].

The S-band frequencies are well suited to the so-called intelligent devices, whose forthcoming exponential growth in the automotive and domotics worlds will benefit from a low-cost high efficient system for non-real-time communications, mainly based on messaging [2]. Telemetry, environment and traffic monitoring, emergency alerts, fleet management, highway tolling, forecast predictions,

P. Pillai, R. Shorey, and E. Ferro (Eds.): PSATS 2012, LNICST 52, pp. 26–34, 2013.

pay-per-view represent just few among all possible applications. Road-safety and emergency applications will be tested in a real scenario through a demonstration platform built-up by SafeTRIP, a collaborative project in the framework of the EU Seventh Framework Program [3].

After a preliminary phase of system definition and performances assessment, carried-out in ESA cofunded projects MiReSys and DENISE, the Enhanced Spread Spectrum Aloha [4] has been chosen as the reference protocol for the return link physical layer specifications, with asynchronous access, and has been recently standardised at ETSI as *Air Interface for S-band Mobile Interactive Multimedia* (S-MIM), Part 3 [5]. The E-SSA structure, based on fully asynchronous random access with a robust Forward Error Correction (FEC) and a simple open loop power control, perfectly copes with low-duty cycle bursty transmission, adapted to satellite messaging. Furthermore, the introduction of ad-hoc Successive Interference Cancellation (SIC) techniques, coupled with a smart reuse of a unique hierarchic preamble, shared by the entire population of terminals, shows good results in terms of MAC throughput. The system capacity allows the simultaneous coexistence of millions of terminals in limited portions of bandwidth, thus lowering the exploitation cost per final user.

This paper presents experimental results based on the first available prototype implementing E-SSA protocol. In Section 2 a system overview is reported, focusing on system parameters. In Section 3 prominent relevance is given to the features emerging from the practical implementation of the protocol, such as the reduced impact of code collision. System capacity results, in terms of maximum allowed MAC-Loads and PLR, are then presented, through different scenarios at increasing standard deviation values in received power and at increasing uniformly distributed frequency errors.

2 System Overview

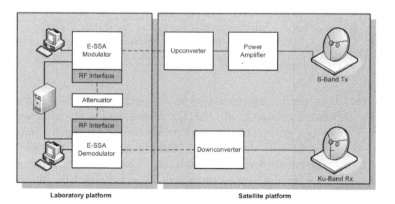

Fig. 1. E-SSA platform at Eutelsat premises

The deployed E-SSA prototype, hosted at Eutelsat premises, is tautly shown in Fig.1. It includes a traffic generator, able to simulate an aggregated signal coming from thousands of terminals simultaneously transmitting over an AWGN channel, and the first available prototype of E-SSA demodulator. Both traffic emulator and demodulator are based on Software Defined Radio (SDR) and are fully reconfigurable to work in a frequency range of 50-2200 MHz. In the present study, the experiences have been run on digital samples generated at the modulator side, stored on disk and then analysed by the demodulator (not in real-time), bypassing the radiofrequency interfaces.

Apart from the laboratory testbed, a complete satellite chain is available at Rambouillet teleport to transmit in S-band towards W2A. First on-air tests have been successfully conducted and preliminary results confirm theoretical expectations. They will constitute the basis of future work.

In Table 1, the most relevant implementation parameters, in line with the S-MIM standard [5], are reported. One of the main differences in system choices, with respect to ESA proposal [6], resides in the increased packet length of 1200 information bits, enabling an increased turbo code efficiency and minimizing the code collision. However, the extended length exposes the packet to channel variations effects within its duration, that may result in additional difficulties at phase and frequency estimation stages.

Table 1. E-SSA implementation parameters

Equipment	Parameters	Value
Modulator	Bandwidth	4,68 MHz
	Chip-rate	3,84 Mcps
	Spreading factor	256
	Number of spreading codes	1
	Info bits per packet	1200
	Coded bits per packet	3600
	Preamble length	96 bits
	CRC length	16 bits
	Oversampling factor	4
Demodulator	Number of SIC iterations	scalable [0-100]
	Max. demodulation attempts	800 packets/s
	Turbo codes iterations	6

Compared to the ideally-continuous packet-by-packet estimation-regeneration-cancellation process envisaged in [6], the SIC mechanism in the analysed implementation applies per groups of detected packets. The amount of packets per group can be scaled at will, defining the maximal number of possible demodulations at each interference cancellation cycle. The SIC mechanism relies on few fundamental operations carried out on each packet. Initially, the preamble searcher allows the detection of the packet. During this stage the chip-timing recovery and the frequency estimation are performed. Afterwards the packet enters its true demodulation stage. The most critical operations are the phase and power estimations. They are performed using a channel estimator that divides each packet in slices and estimates the power and phase of each slice. From this point on, the packet is decoded and CRC checked. If the packet is judged to be correctly demodulated, it is integrated with the rest of the demodulated packets. Considering the previously recovered parameters (chip time, frequency, phase and power), the resulting stream is regenerated and subtracted from the original signal. Thanks to the previous operations, the total interference diminishes and other packets with smaller SNIR arise and are detected by the preamble detector at the following iteration. Fig. 2 well illustrates the iterative SIC behaviour.

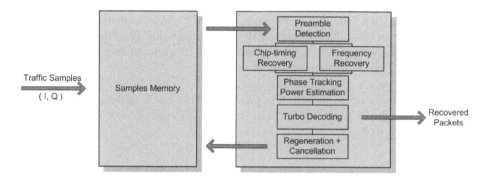

Fig. 2. Successive Interference Cancellation process

3   Performance Analysis

Experimental results coming from the E-SSA implementation allow refining the description of the demodulation process and a better understanding of its crucial steps, i.e. detection, demodulation, channel estimation and cancellation. Thanks to the use of a specific hierarchic preamble of 96 bits, whose choice has been widely motivated in [7], packet detection algorithm can work at increased MAC-Loads. For instance, at 1.28 bits/s/Hz (which correspond to 5000 packets/s, with an occupied bandwidth of 4.68 MHz and a packet length of 1200 information bits), the average E_c/N_t is equal to 30.8 dB. After the preamble correlation phase, the SNIR of the preamble can be calculated:

$$SNIR_{preamble} = \left[\frac{E_c}{N_t}\right]_{th} + 10\log(L) + 10\log(SF) = 13.1\text{dB} \qquad (1)$$

where L is the preamble length in bits equal to 96 and SF is the spreading factor equal to 256, resulting to be a good value for packet detection and SIC effectiveness.

The efficiency of the turbo codes, with the choice of a robust rate ($r=1/3$) results in a lower E_b/N_t threshold for the demodulation, ensuring excellent PLR even at very low packet SNR. Keeping unvaried the assumption made in [6], i.e. packets log-normally distributed in power with a standard deviation of 3 dB, the

Table 2. PLR floors at different burst SNR

Burst SNR$_{average}$ (dB)	PLR floor
-16	$1.3 \ 10^{-4}$
-17	$5.1 \ 10^{-4}$
-18	$1.4 \ 10^{-3}$
-19	$3.8 \ 10^{-3}$
-20	$8.7 \ 10^{-3}$
-21	$1.7 \ 10^{-2}$

system PLR floor may be evaluated. In Table 2, the good PLR values, even at very low SNR, show the system robustness in noisy environments.

Regarding the cancellation process, due to the fact that the efficiency of each stage affects the successive, it is very important to avoid to introduce residual power coming from erroneous cancellations.

The PLR can be also studied at varying MAC-Loads. In order to strictly evaluate the interference cancellation process, no noise was added, resulting in a high SNR. Results in Table 3 confirm the validity of the implemented cancellation process by group of packets.

Table 3. PLR floors for different MAC-Loads

MAC-Load (b/s/Hz)	PLR floor
0.5	$5.8 \ 10^{-6}$
0.7	$7.7 \ 10^{-5}$
1	$4.1 \ 10^{-4}$
1.2	$1.1 \ 10^{-3}$
1.5	$2.7 \ 10^{-3}$
1.8	$5.5 \ 10^{-3}$

3.1   Code Collision

The code collision corresponds to the reception of two or more packets at the gateway demodulator within a preamble detection interval. One can think that the perfect superimposition of packets makes the demodulation unfeasible. An interesting behaviour has been found out during laboratory testing of E-SSA prototype, i.e. the PLR is not lower-bounded by the code collision effect. This is mainly due to the robust rate of the turbo code and to the presence of the SIC. In fact, most of the time the two superimposed packets can be correctly demodulated, even if they have been transmitted at the same chip time. In good channel estimation conditions, if the packets have different amplitudes, it is possible to demodulate and cancel the most powerful of them at a first SIC iteration, then the second one can be found by the preamble searcher at a successive iteration. Similarly, frequency or phase differences can be useful to differentiate between two packets starting at the same time. Moreover, even if

Table 4. Code collision experimental results

MAC-Load (packets/s)	Expected PLR	Measured PLR
1000	$4.2 \ 10^{-4}$	$8.3 \ 10^{-6}$
2000	$0.9 \ 10^{-3}$	$1.7 \ 10^{-5}$
3000	$1.4 \ 10^{-3}$	$1.7 \ 10^{-5}$
4000	$1.9 \ 10^{-3}$	$2.7 \ 10^{-5}$
5000	$2.3 \ 10^{-3}$	$3.8 \ 10^{-5}$

the received packets present almost the same parameters (power, frequency and phase), the probability of good demodulation is still not null.

The explanation comes from the structure of the E-SSA signal, which is obtained by filtering a spread and scrambled BPSK sequence of symbols. From the detection point of view, receiving two different packets aligned at the chip time is equivalent to have a single packet, whose preamble amplitude is doubled, consequently easing the detection. The turbo codes will lead the decoder to choose one of the two initial sequences of information bits. The Table 4 herein confirms this behaviour, showing that collision events do not lower-bound the PLR. In fact, the measured PLR is by far lower than the expected PLR (the ratio between the number of colliding packets and the number of transmitted packets).

3.2   Power Distribution

The burst distribution in received power is one of the crucial system issues. In real scenarios, due to different G/T values within a satellite beam, to the coexistence of several transmitting antennas with distinct gains and to the impairments introduced by the mobile channel, the power received at the hub, even in presence of a quite accurate open loop power control, can vary substantially. This directly affects not only the final performances of the E-SSA demodulator in terms of achievable PLR, but also the number of iterations required to reach the fixed PLR threshold. In order to simulate these effects, a Gaussian power distribution has been chosen.

At increasing standard deviation in log-normally distributed power, two distinct effects are verified from the simulations. First of all, the number of terminals with E_b/N_t lower than the demodulation threshold increases, due to the enlarged basis of Gaussian distribution. Secondly, the SIC algorithm, starting from the cancellation of packets with the best SNIR, is better suited to high power unbalance distributions. In fact, when received power is not spread, the packet detection is much less efficient and the demodulation gets worse, because the average SNR of detected packets is lower. In Fig. 3, it is interesting to notice that up to a normalized MAC-Load of 1.7 b/s/Hz, the system performances are optimized

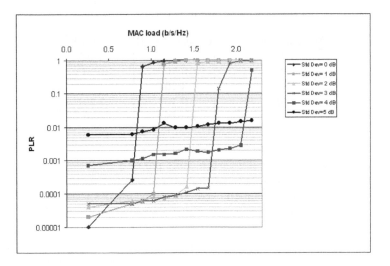

Fig. 3. PLR at increasing MAC-Loads and different values of standard deviation (C/N= -16dB)

at 3 dB of standard deviation both in terms of reached PLR. Above 1.7 b/s/Hz the standard deviation optimal value becomes higher.

3.3   Frequency Distribution

Several components in the E-SSA satellite transmission can introduce a frequency error: the clock instability at the terminal side, the on-board satellite frequency conversion steps, the Doppler effect due to satellite and mobile terminals, the hub frequency accuracy. While the frequency error introduced by the satellite can be compensated at the hub and the hub itself can be easily connected to a stable reference, transmitting terminals conceived for mass market distribution will not embed precise oscillators, as they are still quite expensive. Consequently, the E-SSA demodulator must manage a population of terminals having a certain frequency distribution. Henceforth, frequency errors are assumed to follow a uniform distribution.

In Fig. 4, PLR values are reported for increasing frequency errors of the aggregated traffic. From a general point of view, the demodulator behaves like a selective filter in frequency (-1.5kHz, 1.5kHz), as already demonstrated in [7].

For normalised MAC-Load values up to 1.5 b/s/Hz, the PLR is almost unvaried up to the cut-value of 1500 Hz. Above 1.5 b/s/Hz, the system performances in terms of PLR get worse at increasing frequency errors and, after 1500 Hz, the PLR increases. In other terms, above MAC-Load values of 1.5 b/s/Hz, in presence of frequency drift, the detection process becomes more difficult and higher average SNRs are required to target a fixed PLR.

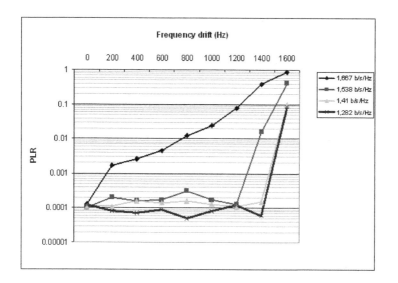

Fig. 4. PLR at increasing frequency error and fixed MAC-Loads (C/N= -16dB)

4   Conclusions

We have shown that the good theoretical performances of the E-SSA protocol
are confirmed by a real prototype. High throughputs at very low PLR (at least
10^{-3}) have been reached and new fundamental behaviours of the system have
been refined. First of all, even if the importance of received packets power unbalance in SIC mechanism had already been stated, here we have shown that to
each MAC-Load corresponds an optimal value of standard deviation in the log-
normally distributed received power. In an enhanced system, a simple open-loop
power control mechanism could exploit this behaviour by signalling from the
hub the ideal received power distribution, as foreseen in the S-MIM standard.
Moreover, it has been shown that the frequency error distribution could be a
constraint especially at high MAC-loads, strongly impacting the resulting PLR.
Future works will deal with the introduction of a mobile channel and the assessment through the radiofrequency interfaces and the end-to-end satellite link.
The channel estimation parameters will become crucial in such an environment
and the impact on performances, jointly with the use of a control channel, will
be studied.

References

1. Arcidiacono, A., Finocchiaro, D., Geurtz, A., Pulvirenti, O., Schluter, G.: S-band:
   a new age in mobile satellite systems. In: 4th Advanced Satellite Mobile Systems
   Conference, Bologna, pp. 122–127 (2008)
2. Arcidiacono, A., Finocchiaro, D., Grazzini, S., Pulvirenti, O.: Perspectives on mobile
   satellite services in S-band. In: 5th Advanced Satellite Mobile Systems Conference,
   Cagliari, pp. 516–521 (2010)

3. Fremont, G., Grazzini, S., Sasse, A., Beeharee, A.: The SafeTRIP project: improving road safety for passenger vehicles using 2-way satellite communications, ITS World Congress 2010, Busan (2010)
4. De Gaudenzi, R., del Río Herrero, O.: Methods, apparatuses and system for asynchronous spread-spectrum communication, European Patent 2 159 926 A1 (2010)
5. ETSI TS 102 721, Satellite Earth Stations and Systems, Air Interface for S-band Mobile Interactive Multimedia, Part 3: Physical Layer Specification, Return Link Asynchronous Access (S-MIM) (2011), http://www.etsi.org
6. De Gaudenzi, R., del Río Herrero, O.: Advances in Random Access Protocols for Satellite Networks. In: International Workshop on Satellite and Space Communications, Siena, pp. 331–336 (2009)
7. Demonstrator Emergency aNd Interactive S-band sErvices (DENISE), Phase 1, Artes 3-4 Projects, Internal Reports, http://telecom.esa.int (2009-2011)

A Redundancy Software Design for Joint Radio Resource Management System in a Satellite-Terrestrial Based Aeronautical Communication Network

Yongqiang Cheng, Kai Xu, and Yim Fun Hu

Schools of Engineering Design and Technology, University of Bradford
Bradford, UK
{y.cheng4,k.j.xu,y.f.hu}@bradford.ac.uk

Abstract. This paper presents a Master/Slave redundancy mechanism for the airborne Integrated Modular Radio to improve the reliability of the joint radio resource management (JRRM) system. The proposed mechanism adopts keep-alive heart beat messages and real time information synchronization to ensure a smooth switchover in the event of a platform failure. To enhance the scalability and decoupling of the system, the proposed hot swap solution makes the JRRM switchover transparent to both the higher layers and the lower layers. The experiment results and the performance obtained from the test-bed has proved the validity of the solution.

Keywords: Aeronautical Networking, RRM Redundancy, Integrated Modular Radio, Master-Slave Switchover.

1 Introduction

The EU Project SANDRA (Seamless Aeronautical Networking through integration of Data-Links, Radios and Antennas) [1] aims to design, specify and develop an integrated aircraft communication system primarily for air traffic management to improve efficiency and cost-effectiveness in service provision by ensuring a high degree of flexibility, scalability, modularity and reconfigurability.

The SANDRA system is a 'system of systems' addressing four levels of integration: Service Integration, Network Integration, Radio Integration and Antenna Integration. From the communications network point of view, SANDRA spans across three segments, namely, the Aircraft segment, the Transport segment and the Ground segment, as shown in Fig. 1. The Aircraft segment consists of three main physical components: the Integrated Router (IR), the Integrated Modular Radio (IMR) and the Antennas. These three components form the SANDRA terminal [2]. While the IR is responsible for upper layer functionalities, such as routing, security, QoS and mobility, the IMR takes care of lower layer radio stacks and functions including radio resource allocation, QoS mapping and adaptation functions. Through Software Defined Radio (SDR) [3] the IMR supports dynamic reconfigurability of operations on a specific radio link at any time and provides the flexibility for accommodation of

P. Pillai, R. Shorey, and E. Ferro (Eds.): PSATS 2012, LNICST 52, pp. 35–43, 2013.

future communication waveforms and protocols by means of software change only. The physical separation between the IR and the IMR has the advantage of increased modularity and identifying distinct management roles and functions for higher layer and lower layer components with IP providing the convergence. The Antennas include a hybrid Ku/L band Integrated Antenna (IA), a VHF antenna and a C-band antenna. The IA is a hybrid Ku/L band SatCom antenna to enable an asymmetric broadband link. The various end-systems i.e. Air Traffic Service (ATS), Airline Operation Centre (AOC), Airline Administrative communication (AAC) and Aeronautical Passenger Communications (APC) [4] are all connected to the IR.

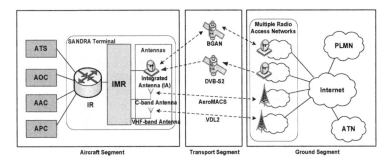

Fig. 1. SANDRA Network Architecture

In the Transport segment, four radio transport technologies are considered, namely, VDL mode 2 [5] in VHF band, BGAN [6] in L-band, DVB-S2 [7] in Ku-band and AeroMACS - a WiMAX [8] equivalence for aeronautics communications - in C-band. The Ground segment consists of multiple Radio Access Networks (RANs) and their corresponding core networks, the Aeronautical Telecommunication Network (ATN), the Internet and the Public Land Mobile Network (PLMN) for passenger communications. In order to provide mobility and security services for aeronautical communications, functional components such as the mobility server, security and authentication server are required in the ground segment to provide corresponding mobility and security information services. These components will be provided by the ATS/AOC/AAC and APC service providers of the ATN on ground.

This paper concentrates on a redundancy design of the JRRM (Joint Radio Resource Management) to increase the reliability of the IMR. Readers are referred to [9] for the functional architecture of the SANDRA terminal for radio resource management (RRM) and an approach to partition the functional entities between the IR and IMR for the configuration and reconfiguration of radio links.

The rest of the paper is organized as follows: Section 2 is the general overview of the JRRM software architecture. Section 3 describes the need for a redundancy approach and the details of the proposed software design to support this redundancy behavior. It also describes the system startup flow procedures. The performance result of the proposed redundancy mechanism is presented in Section 4 and finally Section 5 presents the conclusion of the paper.

2   The SANDRA IMR Software Architecture

Fig. 2 depicts the general software architecture of the SANDRA IMR conforming to the SANDRA network architecture shown in Fig. 1. The hardware platforms are represented by the grey boxes.

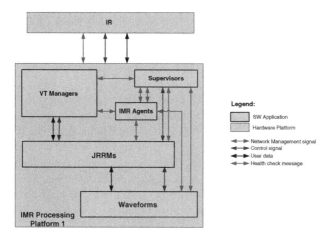

Fig. 2. General Software Architecture of the IMR

On each IMR processing hardware platform, there are five applications running:

- The Supervisor Application: The Supervisor Application is responsible for launching the JRRM and the waveform applications. On boot up, the Supervisor Application will launch the JRRM and tell the JRRM what waveform applications are available on the particular machine. The JRRM can then tell the Supervisor Application to launch a particular waveform. If a waveform application goes down, the Supervisor Application will inform the JRRM. The waveform application might be single application, or maybe a combination of a stack application and a physical layer application.
  The Supervisor Application would monitor all the applications it has launched by sending regular health check messages and monitoring their responses.
- The VT Manager: The Virtual Tunnel (VT) Manager[10] in the IMR works as user plane connection server and also encapsulates the IP packets in the SANDRA specific SAP messages before passing to the JRRM. It is responsible for establishing connections based on the real time information updated by the JRRM.
- The JRRM application: The JRRM, with its provision of adaptation functions and together with the link management functions, forms an adaptation layer between the IR and IMR for interfacing the network layer with the multiple underlying Radio Stacks. This adaptation layer provides the necessary features for the joint RRM purposes. It hides the underlying complexities of the multiple radio protocol stacks from the network layer (i.e. in the IR) and provides a uniform

interface to control the multiple radios. More details of the JRRM internal design and functionalities are defined in [9].

- The Waveform application: Three combinations of the waveforms may be loaded onto one of the two FGPA boards the VDL2+BGAN, the DVB-S2 or the AeroMACS. Thus at a particular time, maximum two combinations can be loaded into the two FPGA boards in the two IMR processing platforms.
- The IMR agent: This is a network management application which is responsible for collecting network management related information for the IMR. It consists of three main parts: a management interface, a Management Information Base (MIB), and the core agent logic.

The JRRM is the centre of all the IMR control plane signals. Control signals are sent to and processed in the JRRM to establish and destroy connections as well as to control behaviours of supervisor and waveforms. User plane data pass through the VT Manager, then the JRRM and reach the waveform for downlink direction and vice versa for the uplink direction data. The IMR Agent make use of management plane signals to collection system status, errors etc. and report all these information to the network management unit. The supervisor periodically sends keep-alive messages to all other applications and monitors the status of them.

3   JRRM Redundancy Design

3.1   Problem Statement

Safety and reliability are essential characteristics of any airborne applications. It has a significant impact upon effective operations. Although at present it may not be an essential requirement for the SANDRA IMR prototype to gain a DO-178B certifiable status [11], it is necessary for a commercial IMR to obtain the DO-178B certification in the future. As the key intelligent sub-component of IMR, the JRRM is critical for the normal operation of the whole SANDRA communication system to forward packets, resolve address, map QoS, manage links and etc. A failure of the JRRM will result in the failure of the IMR and thereby the whole SANDRA system. Hence to avoid such a failure it is imperative that the JRRM should backup all live session information and have the capability to switch all connections to the backup JRRM to provide uninterrupted service.

Also in order maintain the scalability of the IMR system, it is important that the redundancy mechanism is completely transparent to the supervisor and the underlying waveform protocol stacks.

The redundancy mechanism should be also transparent to the higher layers in the IR to minimize the coupling of these two modules and ensuring the modularity of the system. Another additional requirement for aeronautical communications is that the Master and Slave Node should switch their roles per flight base.

3.2   JRRM Redundancy Design

A 1+1 (Master/Slave) level redundancy solution is adopted for the JRRM to increase the reliability of the IMR so as to reduce the system failure level. As depicted in Fig. 3, the IMR consists of two IMR processing hardware platforms. Both IMR processing platforms have the same configurations and applications running except that the JRRM A in the IMR processing platform 1 works in a Master mode while the JRRM B in the IMR processing platform 2 works in a Slave mode. A Master mode node will be responsible for handling all user traffic and control signals from/to the IR and other applications; a Slave mode node only synchronizes information with the Master. The VT Manager co-located with the Master node is always active and process traffics while the one co-located with the Slave node synchronizes with the Master.

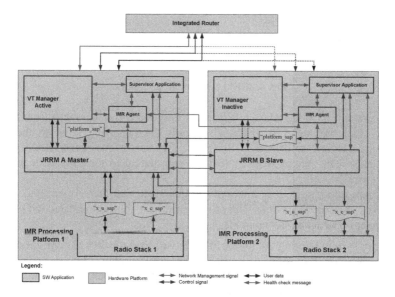

Fig. 3. The Architecture of JRRM Redundancy Design

Keep-alive messages are sent by the Master JRRM to the Slave JRRM and should be responded by the Slave JRRM which ensures that both the Master and the Slave JRRMs know each others status. Once the Slave JRRM detects the failure of the Master JRRM, it will take over the task of the Master JRRM immediately and becomes the Mater. Both the Master and Slave JRRM nodes are configured with the same IP and MAC addresses for the interface towards the IR while one being enabled and the other one being disabled. From the IR's point of view, only one IMR processing platform exists.

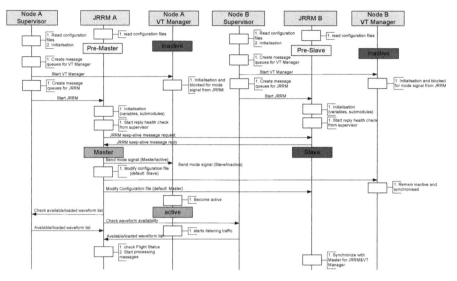

Fig. 4. JRRM system start up routines

As shown in Fig. 3, the Master JRRM not only reads the messages from the local supervisor and the waveforms, but also reads the ones belonging to the remote applications. In the case of Master-Slave role switchover, the Slave JRRM will read all the queues. This makes JRRM's location transparent to the supervisor as well as the waveforms.

On system booting up, it is important to guarantee that there is one and only one Master exist at a particular time. This is achieved by a confliction detection procedure during system boot up as shown in Fig. 4.

1. On system booting up, all the JRRMs will read the JRRM configuration files stored on the local disk. From the configuration file, it is assigned a preliminary working model either as pre-Master or pre-Slave. This will enable the two JRRMs to switch working mode per flight base rather than randomly picked up status.

2. The JRRM will initiate all sub-modules as the pre-assigned working mode.

3. The JRRM assigned with Master mode starts sending keep-alive message to the Slave JRRM. The slave JRRM starts listening and responding to the keep-alive message. If the Master JRRM receives keep-alive messages request from another Master, it will compare the system up time with the one contained in the keep-alive message, the one starts first will be the winner. Similarly, it a Slave JRRM has not received keep-alive message for a certain period of time, it will starts sending keep-alive message request and check whether another Master exist and decide whether to take the role of Master.

4. The Master JRRM selected from the above step will be the formal active Master and be responsible to rewrite the configuration files on both platforms to swap the predefined master/slave modes. Mode change indications will also be sent by the master JRRM to all VT Managers to work as master or slave.

5. The Master JRRM performs its role, such as sending keep-alive message requests, handling all control signals and user data, checking flight status to load waveforms etc..

4   Performance Evaluation and Results

The test-bed developed to measure the performance of the proposed mechanism is consists of an IR which is a Dell Vostro 430, Interl® Core™ i5 CPU, 2x2.67GHz, 2.00GB Memory PC running Fedora Linux and two IMR Processing platforms with the same model PCs as IR but running QNX 6.5.0.

The results in Fig. 5 shows the user plane data round trip delays before and after the Master-Slave switch over. It is assumed that the waveform emulator and the Master JRRM are on the same IMR processing platform before Master-Slave switch. As a consequence of the Master-Slave switch, the Slave JRRM becomes the Master and it is running on a different IMR processing platform from the waveform (cross platforms scenario). For one packet size test, the delay is measured as follows,

1. The IR stamps the system time and sends a packet to the JRRM;

2. The JRRM process the data and forward the data to the waveform emulators;

3. The waveforms emulators swap the Destination and source address of the packet and echo the same payload to the JRRM;

4. The JRRM analyze and return the data back to the IR;

5. The IR stamps the system time again and compare the time difference.

6. Repeat step 1-5 for one packet size 2000 times and get an average of one packet size to avoid random system error.

It was seen that as the packet size increases from 0 to 1400 bytes, the delay increases from 1.8 milliseconds to 2.8 milliseconds. The delays increases after the Master-Slave switchover, because the waveforms and the JRRM are on different hardware platforms and the cross-platform transmission of the data takes extra time, around 2 milliseconds.

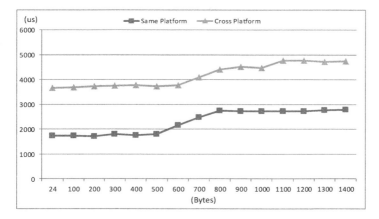

Fig. 5. Round trip delay

Table 1 lists the time consumed by different modules during the Master-Slave switch process. Regardless the number of active connections being established, health check module and traffic processing module consumed 3.0 milliseconds and 6.0 milliseconds respectively to become a fully functional Master from Slave mode. The time for the VT Manager to function from inactive to active varies from 5.0 milliseconds to 12.0 milliseconds depend on the number of active connections.

Table 1. Master-Slave Switch over time

Modules / Active Connections	1	5	10
Health check module	3.0 ms	3.0 ms	3.0 ms
Processing module	6.0 ms	6.1 ms	6.1 ms
VT Manager	5.0 ms	8.5 ms	12.0 ms

5 Conclusion

This paper presents the Master-Slave redundancy mechanism for the IMR. The design, signaling procedures and SAPs involved in the backup solution are described in detail. The solution also meets the requirement of the transparency of the JRRM switchover towards other components, such as the IR and the waveforms. This approach is validated and proved to be efficiency by the experiments results. The overall delay within the IMR system is far less than 10ms no matter the system is in normal operation or post-switched across platforms operation.

Acknowledgment. The research leading to these results has been partially funded by the European Community's Seventh Framework Programme (FP7/2007-2013) under Grant Agreement No. 233679. The SANDRA project is a Large Scale Integrating Project for the FP7 Topic AAT.2008.4.4.2 (Integrated approach to network centric aircraft communications for global aircraft operations).

References

[1] http://www.sandra.aero
[2] Xu, K., Pillai, P., Hu, Y.F., Ali, M.: Interoperability Among Heterogeneous Networks for Future Aeronautical Communications. In: S.P. (ed.) Future Aeronautical Communications, InTech (2011)
[3] Reconfigurable Radio System (RRS). Software Defined Radio Reference Architecture for Mobile Device, ETSI TR 102 680 V1.1.1 (March 2009)
[4] Long-Term Forecast, Flight Movements (2006 - 2025). EUROCONTROL STATFOR (December 2006)
[5] Manual on VHF Digital Link (VDL) Mode 2. ICAO Doc 9776 AN/970 (2001)
[6] http://www.globalcoms.com/products_satellite_bgan.asp

[7] Digital Video broadcasting (DVB) 2nd generation framing structure, channel coding & modulation systems for broadcasting, interactive services, news gathering and other broadbands satellite applications (DVB-S2). ETSI EN 302 307 V1.2.1 (August 2009)

[8] Local and metropolitan area networks Part 16: Air Interface for Broadband Wireless Access Systems (Revision of IEEE Std 802.16-2004), in IEEE Std. 802.16 (2009)

[9] Ali, M., Xu, K., Pillai, P., Hu, Y.F.: Common RRM in Satellite-Terrestrial Based Aeronautical Communication Networks. In: Giambene, G., Sacchi, C. (eds.) PSATS 2011. LNCS, Social Informatics and Telecommunications Engineering, vol. 71, pp. 328–341. Springer, Heidelberg (2011)

[10] Hu, Y.F.: SANDRA_D4.3.5.1_Resource and Link Management Design Report and Validation Test Plan (2011)

[11] Baddoo, J.: SANDRA_IMR_PropoyeOptions_RTOS. SANDRA Working Document (2010)

Spectrum Availability for Next Generation Satellite Services: Coexistence with Terrestrial Mobile Services

Henk Dekker, Bram van den Ende, Hugo Gelevert, and Peter Trommelen

Netherlands Organization for Applied Scientific Research (TNO)
{henk.dekker,bram.vandenende,hugo.gelevert,
peter.trommelen}@tno.nl

Abstract. Having available adequate and sufficient spectrum resources is a crucial factor to enable the fast growth of broadband mobile communications. Efficient use of scarce radio spectrum becomes more and more important. Especially in the lower frequency bands, with favorable conditions for mobile communications, spectrum will have to be shared between different services and applications. In this context effective sharing between mobile service and satellite services becomes increasingly important. It is no longer affordable to base the interference calculations on worst case assumptions. Therefore a novel and comprehensive approach of coexistence analysis is presented. The method discusses the extent to which propagation paths of various interference sources are correlated or not, and what the expected effect will be. Having a more realistic insight in the interference conditions could provide better viability of sharing arrangements, possibly with some realistic mitigation measures.

Keywords: satellite services, MSS, FSS, mobile communication, mobile service, MS, broadband, spectrum, sharing, propagation, ITU-R P.452, model, coexistence, correlation radio path losses, interference analysis.

1   Expected Demand for Mobile Services

It is commonly recognized that terrestrial mobile communications have shown tremendous growth over the last decades. Building on this success the technology evolved to subsequent generations of terrestrial mobile networks, called IMT (International Mobile Telecommunications) systems, providing broadband wireless data services that support novel applications, such as internet browsing, e-mail, messaging, social media and video sharing.

The enormous number of mobile users as well as the increase of mobile broadband applications boosts the need for wireless transmission capacity and as a consequence radio spectrum. The industry is responding to these needs by developing new technologies with a vastly enhanced performance (e.g. IMT-Advanced) and a more efficient and flexible use of the radio spectrum. The future use of the most advanced mobile technologies will provide significant gains in spectrum efficiency but will not be sufficient to cope with the growing demand for capacity. Studies from the International Telecommunications Union (ITU) [1] predict that the total spectrum

P. Pillai, R. Shorey, and E. Ferro (Eds.): PSATS 2012, LNICST 52, pp. 44–52, 2013.
© Institute for Computer Sciences, Social Informatics and Telecommunications Engineering 2013

requirement for mobile cellular systems in the year 2020 will be significantly higher than the total frequency bands identified for the terrestrial component of IMT. Since adequate spectrum availability is considered to be a prerequisite for the success of the continuing development of IMT, there is a huge pressure to more make spectrum available for this purpose.

At the same time, the use of Mobile Satellite Services (MSS) is also expected to increase significantly over the next decade. This growth is driven by demand, for MSS offers (global) applications for handheld mobile terminals and is very well suited for the delivery of among others broadband data and internet services in rural areas. For this reason, MSS is seen as an important element of the European Digital Agenda, a key goal of which is to ensure broadband access for all European Citizens. But MSS also makes possible new applications and services, like for example asset tracking and fleet management in the maritime and transport sector, onboard communication services offered to airline passengers, monitoring of remote pipelines and oil installations and machine-to-machine applications in various sectors. Moreover, MSS offers a solution for emergency communications in regions where the fixed infrastructure has collapsed as a result of a natural disaster or a crisis. These so-called 'hot spots' can be reached by new satellites that are equipped with steerable spot beams. As a result of the above, the number of MSS handheld terminals in service is expected to be doubled or even tripled in 2020 and the revenues associated with MSS are predicted to grow at a rate of 7 to 8 percent per annum [2] [3], as figure 1 illustrates. Major satellite operators confirm these growth expectations, although some reservations are being made for the short term [4].

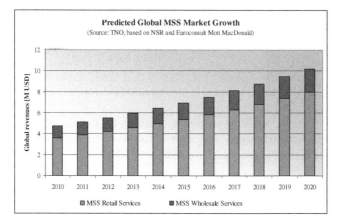

Fig. 1. Predicted global MSS market growth until 2020

Serving the increasing number of MSS users and fulfilling the demand for enhanced broadband services presents a challenge to satellite operators. Having guaranteed and undisturbed access to sufficient spectrum resources is an essential basis for the successful operation of a satellite network and provisioning of reliable services. Since mobile communications are expected to grow, the demand for spectrum is increasing. A consequence of the expected increase in the use of MSS is that the feeder links will also require more capacity which implies the need for sufficient spectrum for Fixed Satellite Services (FSS) as well.

2   Spectrum Requirements

In general MSS and Mobile Services (MS) are seen as complementary services, as is illustrated in figure 2. In the IMT concept a satellite component has been included from the beginning onwards to provide voice and data communication services in regions outside terrestrial coverage. From the satellite service point of view a Complementary Ground Component (CGC) is envisaged to provide mobile satellite or broadcasting satellite services in areas where satellite reception is difficult. CGC consists of base station type equipment that may be collocated with terrestrial cell sites or can be placed stand-alone. To enable compatibility and easy terminal hand-over, terrestrial and satellite radio interfaces are required to have a high degree of common functionality and to operate in the same or adjacent frequency bands.

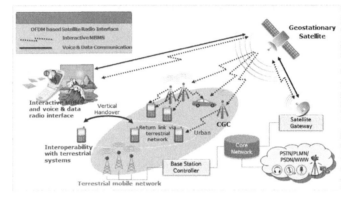

Fig. 2. MSS as a component in the overall IMT-Advanced concept [5]

To enable portable satellite terminals, similar to mobile phones used in the terrestrial mobile networks today, operation in the lower frequency bands (L, S and C-band) with their favorable propagation conditions is desired.

3   Frequency Allocations

During the consecutive ITU World Radio Conferences (WRCs) additional frequency bands were step by step allocated to MS, as shown in figure 3. The allocations for MS will affect the spectrum that is available for satellite services. This is the case in the band around 2 GHz where spectrum allocated to MSS is intertwined with the MS allocations. In the 3600 to 3800 MHz band there is an allocation for both FSS (i.e. feeder links for MSS systems) as well as MS, which means frequency sharing between both services becomes important. The trend is that the spectrum allocations for satellite services are under pressure and are not likely to increase.

Migration of systems that might as well use higher frequency bands (Ku or Ka band), such as VSAT, can make the lower bands more available for portable satellite terminals. This measure is not expected to be sufficient and the need for sharing between MS and MSS/FSS will become increasingly important.

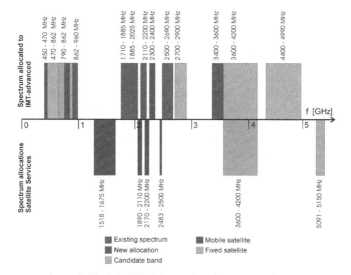

Fig. 3. Spectrum made available for IMT-Advanced and spectrum allocations for FSS and MSS (mostly on a shared basis)

4   Satellite Sharing with Terrestrial Mobile

In case of wireless communication networks with terrestrial and/or satellite components, operated and controlled by a single provider, spectrum sharing reduces to an optimization issue. However, in case different providers are involved the situation is more complex and conflicting requirements have to be dealt with, like interference criteria, mitigation techniques to improve spectrum sharing and the cost involved with implementing the mitigation measures. There is an urgent requirement to optimally exploit the possibilities for sharing and coexistence between radio services. Against this background, there is a need to perform the coexistence analysis as accurately as possible. This paper discusses a novel and comprehensive approach to further improve the coexistence analysis between FSS and MS services in the 3400 to 3600 MHz band, a case that is of particular current interest.

For MS not to interfere with existing FSS links, two criteria are taken into account [6]: the long-term and short-term interference criterion. The long-term interference criterion is used to ensure a good quality of the satellite link and specifies the interference level that must not be exceeded for more than 20% of the time. For satellite earth stations in the fixed satellite service this interference level amounts to 10% of the clear sky satellite system noise. For this criterion the interference contributions from all interference sources are assumed to vary simultaneously and are assumed to add on a power basis, meaning that interference contributions (and path losses) are considered to be correlated.

The short-term criterion is used to ensure the availability of the satellite link and specifies the maximum time percentage that a satellite link may be out-of-service due to a high level of interference (i.e. causing an increase of the satellite system noise which exceeds the satellite link margin). For this criterion the interference contributions from all interfering sources are assumed to vary independently and to add on a percentage-of-the-time basis, meaning that interference contributions (and path losses) are considered to be uncorrelated. For analogue satellite links this outage percentage is 0.03% and for digital links it is 0.005% [6].

To determine whether or not the interference criteria are exceeded, the propagation model described in ITU-R P452 [7] is used. We implemented this model and terrain data was used to calculate the interference level caused by a MS base station at a satellite earth station for various time percentages. Subsequently, for any (range of) azimuth and elevation angles of the satellite ground station, the percentage of time that an interference criterion is exceeded is calculated.

The example illustrated in figure 4 shows the areas around a satellite earth station within which a single MS base station will cause the long and short term criterion to be exceeded. Obviously, the area within which a single MS base station could cause interference exceeding the short term criterion is much larger then for the long term criterion. To obtain an area comparable to that of the long term, either a high outage percentage (of about 5%) has to be accepted or a high satellite link margin (of about 26 dB instead of 3 dB) has to be used by the FSS operator. Clearly, the short term criterion has more impact on spectrum sharing possibilities than the long term criterion. In the following analysis the focus will therefore be on the short term criterion.

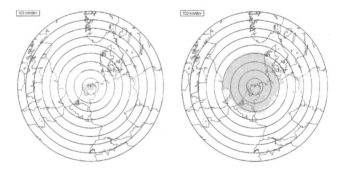

Fig. 4. Example of areas within which the long term (left) respectively short term (right) criterion will be exceeded by a single MS base station

In general the assumptions made as basis for the criteria and the way in which the total interference caused by multiple MS base stations at a satellite earth station is calculated, are well suited for the case that only a limited number of dispersed MS base stations have to be considered.

However, in case a large scale deployment in cities is considered, with many MS base stations in a clutter environment, two complications arise. Firstly, the calculation becomes rather complex and a large amount of environmental data is required to determine the clutter loss for each base station. Secondly, the assumption made for the short term criterion that the interference contributions of all MS base stations arriving at the satellite earth station are uncorrelated becomes unlikely. Considering for instance a large scale MS network deployment in a city, at a large distance from a satellite earth station, the propagation paths between each MS base station and the satellite earth station are very close together (nearly identical from a propagation perspective). This suggests that the interference contributions of the MS base stations at the satellite earth station are more likely to be highly correlated even for short periods of time. This means that the periods of high interference at the satellite earth station due to each base station will occur more simultaneously instead of randomly spread in time. As a consequence, the actual outage percentage caused by a MS network deployment in a city can be considerably less than predicted when considering the contributions of all interferers uncorrelated as is currently common practice.

An alternative approach is suggested here, which is dedicated to modelling the interference effects of a large scale MS network deployments in cities on FSS systems. This approach has been used to estimate the outage percentage at satellite earth stations in the Netherlands for various scenarios (i.e. MS network deployments in various cities).

To reduce the computational complexity, a large scale MS network deployment in a city is modeled as a single (equivalent) base station by assuming propagation losses between each individual MS base station and the interfered-with satellite earth station are highly correlated. For this, a real large scale deployment of a broadband mobile network (based on Wimax technology) in Amsterdam served as a reference. For this network deployment the total power per channel, radiated towards the horizon, was measured. These measurements are subsequently used to determine the E-EIRP (Equivalent - Effective Isotropic Radiated Power) of a single base station in an uncluttered environment resulting in the same total power per channel towards the horizon (see figure 5). Knowing all the specifics of the BWA deployment in Amsterdam, the deployments in other cities are modeled in the same manner by scaling the E-EIRP according to the corresponding coverage area.

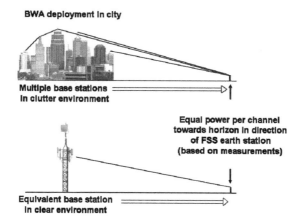

Fig. 5. Modelling of MS deployment in a city

The total outage percentage due to deployments in several cities still depends on the extent to which the propagation loss on paths between the base stations in the various cities and the satellite earth stations are correlated. Again these paths can be close together or partly coincide (suggesting some or high correlation) or be widely separated (suggesting low or no correlation). At the moment that this study was performed, no model to predict the correlation in the path losses between different paths was available. This forced us to compute both the best-case (full correlation) and worst-case (no correlation) situation which, even for a limited number of cities with MS network deployments, leads to calculated outage percentages with a large difference.

As an example, we consider the case of a large scale MS network deployment in 15 cities (of which one is in Germany) and its effect on FSS systems located in Burum.

Fig. 6. Example of a large scale MS network deployment in 15 cities

Since in Burum multiple satellite ground terminals are present, the outage due to MS interference has been calculated for all visible satellite locations. For the satellite ground terminals, the following parameters are used: 11 meter dish, 75°K system noise temperature. In addition, the (satellite) link margin is taken to be 3 dB.

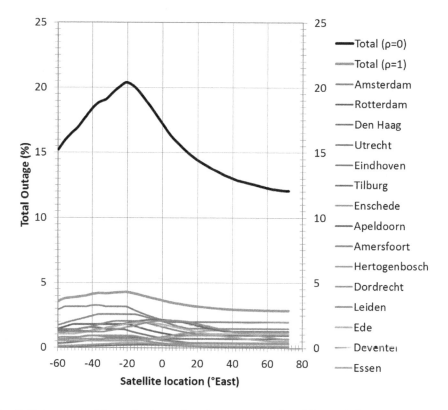

Fig. 7. Outage caused by each city and total outage assuming full correlation (ρ=1) and no correlation (ρ=0)

As shown in the figure above, the large scale MS network deployment results in outages of 2.9 ~ 4.3 % (correlated) and 12.1 ~ 20.4 % (uncorrelated) respectively, depending on which satellite (location) the satellite ground terminal is aimed. This shows that both assumptions (fully correlated or uncorrelated) lead to quite different results.

To obtain more accurate predictions, a recognized model to predict the correlation of the path losses between different paths is highly desired. Although not yet available, research is on-going in order to gain further insight in the correlation between path losses and thus to be able to develop an adequate model [8], [9].

5   Conclusion

Spectrum sharing requires a model by which interference can be accurately predicted. The current model is suitable if only a small number of dispersed interferers have to be considered. For more complex scenarios, as cellular like deployments in cities where many base stations are located within a relative small area in a clutter environment, it is less suited and will result in an overestimation of the outage percentage. To obtain more accurate predictions, a recognized model to predict the correlation of the path losses between different paths is required. This requirement is stressed by the burdens involved with implementing the mitigation techniques necessary to meet the objectives, which should be kept to a minimum.

References

1. ITU-R Report M.2078: Estimated spectrum bandwidth requirements for the future development of IMT-2000 and IMT-Advanced (2006)
2. Euroconsult Report Forecasts 7 Percent Annual Increas. In MSS Revenue Over Next Decade (2011), http://www.spacemart.com
3. NSR: Mobile Satellite Services Market Forecast to Reach $10.2 Billion by 2020. Press release (September 9, 2011)
4. Inmarsat, Inmarsat plc Reports Interim Results 2011. Press release (August 8, 2012)
5. ITU: Regulation of Global Broadband Satellite Communications (GSR Advanced Copy) (2011)
6. ITU-R Recommendation SF.1006: Determination of the interference potential between earth stations of the Fixed-Satellite service and stations in the Fixed Service (1993)
7. ITU-R Recommendation P.452-13: Prediction procedure for the evaluation of microwave interference between stations on the surface of the earth at frequencies above about 0.7 GHz (2007)
8. Craig, K.H.: Theoretical Assessment of the Impact of the Correlation of Signal Enhancements on Area-to-Point Interference. Rutherford Appleton Laboratory, Oxfordshire (2004)
9. Willis, M.J., Craig, K.H.: Results from a Long Term Propagation Measurement Campaign. Antennas and Propagation (2007)

Interference Analysis Due to Spot Beams Drift in Integrated Satellite-Terrestrial Networks

Helal Chowdhury and Janne Lehtomäki

Center for Wireless Communications, University of Oulu, Finland
fougi@ee.oulu.fi

Abstract. Integrated satellite-terrestrial networks can improve both the service coverage and spectral efficiency in spot-beam-based mobile satellite service system (MSS). In an integrated satellite-terrestrial network, MSS frequency is reused in both satellite spot beams and complementary ground component (CGC). Due to reuse of frequency, co-channel interference is present in both intra-component (between CGCs) and inter-component (between satellite spot beam and CGC). Without spot beams drift, these interferences have tolerable levels for acceptable performance in voice and/or data communication. However, in practice, especially the inter component interference will start to grow if the spot beams drift from their nominal positions due to pointing error of spot beams. In this paper, we study the interference from the satellite component to the ground component when the drifted spot beams start to interfere with CGCs reusing the same frequency. The interference levels depend on the drift magnitude, drift angle, and also the protection ring size utilized by the ground component. Simulation results are provided for different values of these parameters and giving new insight into the effects of spot beam drifting in an integrated satellite-terrestrial network.

1 Introduction

Recently, the International Telecommunication Union (ITU) has defined the concept of an integrated mobile satellite service (MSS) system [2,3,4,6]. In this concept both satellite based network (SBN) and complementary terrestrial based network (CTN) are interconnected and controlled by satellite resource and network management system [3].

CGC is ground-based component which interact directly to the core network [6]. CGC is also called by CGC type 1. Type 1 is L3 type of relay where from user equipment (UE) point of view is considered as a cell of its own. Different types of CGCs are discussed in [6]. In this paper (here after), we will use the terms CGC and terrestrial cell interchangeably. In both of these entities MSS band is used to provide services seamlessly directly by satellite or via CGC. The main purpose of building such integrated satellite-terrestrial network is to improve satellite service coverage in areas where the satellite communications suffer from high blocking factor. On the other hand, spectral efficiency is increased by reusing frequency in both SBN and CTN [3].

However, having benefits of increasing spectral efficiency and improving service coverage using integrated satellite-terrestrial system, therein remain significant challenges to be resolved to foresee the successful deployment of integrated satellite-terrestrial system. One of the major challenges is to avoid or mitigate intra- and

P. Pillai, R. Shorey, and E. Ferro (Eds.): PSATS 2012, LNICST 52, pp. 53–60, 2013.
© Institute for Computer Sciences, Social Informatics and Telecommunications Engineering 2013

inter-component co-channel interference caused by using same frequency in integrated terrestrial-satellite system. Intra-component interference is generated within terrestrial cells. On the other hand, inter-component interference is generated between terrestrial cells and satellite spot beams. In ideal situation these interferences have tolerable levels for acceptable performance in voice and/or data communication. But in practice the spot beams are not fixed to the Earth surface. There are various types of natural perturbing forces which will cause the GEO (Geostationary Earth Orbit) satellite to drift out from its original path and assigned position towards so-called inclined orbit [5]. Hence, antenna system will not be anymore aimed properly with the right pointing of its spot beams projection towards the Earth surface. As a result, inter component co-channel interference will start to grow if the spot beams drift from their nominal positions.

Very few papers have discussed the co-channel interference issues in integrated satellite-terrestrial systems [1,8]. In [1], a model of an integrated mobile network composed of a multiple spot beams satellite and terrestrial base stations (BSs) is presented. This model is then used for analyzing uplink co-channel interference issues in integrated satellite and terrestrial mobile systems. Moreover, analysis of interference issues in integrated satellite and terrestrial mobile system have been done by assuming that satellite spot beam is completely fixed relative to the Earth surface. In [8], experimental based study has been conducted to investigate the interference issues in integrated satellite-terrestrial system. In both [1] and [8], drift of spot beams are assumed to be maintained fixed. However, in inter-component interference analysis the knowledge of drifting of spot beam is essential. Depending on the pointing accuracy of the spot beams, the position of the satellite spot beam pattern compared to the terrestrial cells can vary within a range of several of kilometers. Hence, performance analysis of integrated satellite-terrestrial system under the influence of spot beam drift is an important issue to study.

In this paper, we will analyze both intra component and inter component downlink interference issues for integrated satellite-terrestrial system when the multiple satellite spot beams drift. In our usage scenario, multiple satellite spot beams cover a large geographic area which could include a large number terrestrial cells. Thus, when the adjacent spot beams drift, satellite downlink interference could impinge on a large group of co-channel terrestrial cells within the scope of satellite spot beams drift. The interference level depends on the drift magnitude, drift angle and also on protection ring size around the spot beam.

2 System Model

For co-existence assessment, in this study, GEO (Geostationary Earth Orbit) multibeam satellite is considered as space component and CGC as ground component. Geographic target area is considered as rectangular area. The target area is then sub-divided into many service areas where services will be provided to users. For the sake of simplicity we will consider only one service area which resides in the middle of the target area. A cluster of spot beams of integrated satellite-terrestrial system and its overlaid CGC cells is shown in Fig. 1. The coverage of the spot beam is modeled by two geometrical shapes. Nominal coverage is bounded by hexagonal shape and the outer coverage of the spot beam is bounded by circular shape. This circular shape ring is called protection ring. Protection ring is basically part of spot beam coverage. The purpose of

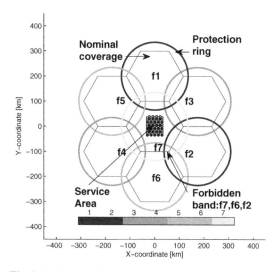

Fig. 1. A cluster of spot beams (7 spot beams in one cluster)

creating protection ring around each satellite spot beam is for not emitting or receiving too much interference to/from adjacent spot beams and terrestrial cells. Inside each protection ring the frequency used by the corresponding spot beam is not allowed to be reused by the CGC. An example of forbidden channels within the cross section of spot beams is also shown in Fig. 1. Very large size of protection ring will reduce inter-component interference. However, it will lead to a reduced set of MSS channels left available to the CGC [1]. Transmission power for each CGC cell is same. Frequency allocation in each spot beam and terrestrial cells for a specific service area is done in such a way that they are orthogonal. The shape of terrestrial cells is also considered as hexagonal. Cell size varies depending on the environment such as rural and urban case. Nominal positions of all spot beams in ideal situation (without drift) is shown in Fig. 1. Users distribution in every terrestrial cell is assumed to be uniformly distributed. There are several ways to plan the reuse pattern for terrestrial networks. In this paper, we will use frequency reuse factor 3 for terrestrial cells and 7 for satellite spot beams. If the given MSS bandwidth is 50 MHz then for a particular spot beam and its overlaid terrestrial cells, MSS band will be divided equally into seven sub bands. One sub band

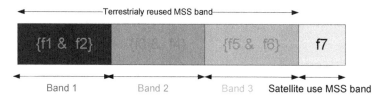

Fig. 2. MSS band allocation in satellite spot beam and terrestrial cells (14.2 MHz/CGC and 7.1 MHz/spot beam)

will be allocated to satellite which is equivalent to 7.1 MHz and the rest will be used for terrestrial cells. In this case, each sub band will be equivalent to 7.1 MHz. However, due to our assumption of frequency reuse factor 3 for terrestrial cells, allocated six sub bands will be combined to form three bands. Hence, each band of terrestrial cell will be basically combination of two sub bands which is equivalent to 14.2 MHz. Band allocation to a particular terrestrial cell is shown in Fig. 2 where blue, green and cyan are represented as terrestrially reused MSS band and yellow is represented as satellite spot beam used band (here f_7). Moreover, in Fig. 2, blue color band is termed as band 1 and is composed of two sub bands $\{f_1, f_2\}$. Similarly, band 2 (green) and band 3 (cyan) are composed of $\{f_3, f_4\}$ and $\{f_5, f_6\}$ respectively. Frequency plan is carried out in C band either between 3.4 GHz and 3.55 GHz or between 3.6 GHz and 3.75 GHz.

3   Drift of Spot Beams

The pointing error of drifted spot beam may vary depending on the orbit chosen for satellite. The statistics of the drift trajectory can be obtained via measurement. However, in this paper, we use a simple pointing error model to analyze the interference issues under the influence of multi spot beams drift. In modeling pointing error of drifted spot beam, we characterize it by two key parameters: drift angle and drift magnitude. Drift angle will define the direction of spot beam drift. On the other hand, drift magnitude defines length of trajectory path towards a certain direction of spot beam drift. Unless otherwise specified drift angles are uniformly distributed. However, in this work most results assumed given fixed drift angles.

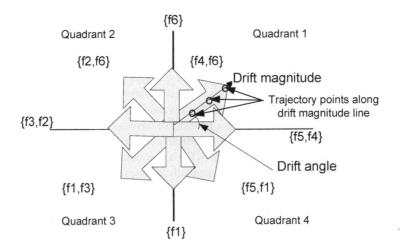

Fig. 3. Orientation of inclined and non-inclined drift angles of spot beams

According to the orientation of drift angle we can characterize it into two types of pattern of drift angle: non-inclined and inclined. The details of two different orientation of drift angles are given below:

In inclined orientation, drift angle of spot beams will be inclined to any one of the four quadrants as shown in Fig. 3. For example if the drift angle resides in quadrant 1 then two spot beams from the opposite quadrant will interfere the cells, in this case $\{f_4, f_6\}$ will be responsible spot beams to interfere the target service area. Similarly if the drift resides in quadrant 2, quadrant 3 and quadrant 4 then the set of spot beams such as $\{f_2, f_6\}$, $\{f_1, f_3\}$, $\{f_1, f_5\}$ will be responsible to interfere with the service area respectively.

In non-inclined orientation of drift angle, spot beams will drift either horizontally or vertically. Moreover if it drifts horizontally it can drift either left side or right side from its nominal positions. In this case $\{f_5, f_4\}$ and $\{f_3, f_2\}$ will be responsible sets of spot beams to interfere with the service area. On the other hand, if it moves vertically it can drift either top or downwards from its nominal position. In this case, only one spot beam either $\{f_1\}$ or $\{f_6\}$ will interfere with the service area. However, all of the above considerations are for moderate realistic protection ring sizes.

4   Channel Model

The path loss model used in this work is [7]:

$$A = A_0 + 10\gamma \log\left(\frac{d}{d_0}\right) \tag{1}$$

where A_0 is the intercept attenuation at a distance d_0 from the CGC, γ is the pathloss exponent. d_0 and d are the reference distance and distance between user and CGC. The signal-to-interference ratio (SIR) in terrestrial system depends on many factors, including the cell layout, cell size, reuse distance, and propagation. SIR for a non-collided user is

$$SIR = \frac{P_{desired}}{\sum P_{interference}} \tag{2}$$

where $P_{desired}$ is the signal strength from the desired CGC and $P_{interference}$ is the interference signal strengths from neighboring co-channel CGCs. The SIR for a collided (victim) user contains both the co-channel interference and inter-component interference from the overlapping drifted satellite spot beam and is written as

$$SIR = \frac{P_{desired}}{\sum P_{interference} + \sum P_{satellite}} \tag{3}$$

where $P_{satellite}$ is the summation of satellite interference powers experienced by the desired user from the co-channel drifted spot beams.

5   Simulation Results

In this section interference of integrated satellite-terrestrial has been evaluated using pointing error model presented in Section III and the parameters presented in Table 1. In our simulation, we consider geographic area of 70×70 km^2 reflecting the size of a

Table 1. Simulation Model Parameters

Simulation Parameters	Value
Service Area	$70 \times 70 \text{ km}^2$
Protection ring size	135 km
Frequency reuse factor for cellular	3
Frequency reuse factor for satellite	7
CGC cell radius	5 km
Drift angle	Uniform or $(0^o, 45^o, 90^o)$
Number of users per cell (CGC)	100
Path loss exponent (α)	3.8
A_0	105.45 dB
Reference distance (d_0)	200 m
Satellite interference power	-104 dBm
Transmission power (CGC)	43 dBm
Minimum receiver sensitivity (CGC)	6.4 dB

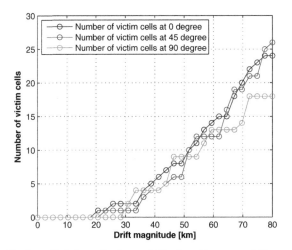

Fig. 4. Number of victim cells due to drift of spot beam vs. drift magnitude

service area of a city. As performance metrics, we use the number of victim cells and received SIR level.

Fig. 4 shows the number of victim cells with respect to the different length of drift magnitude of the spot beam(s). The studied drift angles were 0^0, 45° and 90°. It can be seen that with 135 km protection ring radius and for all chosen drift angles there are no victim terrestrial cells up to 19 km of drift magnitude of spot beam. However, number of victim cells will start to grow after 20 km. It is also seen that when drifted spot beam moves from one trajectory point to another trajectory point, the pattern of victim cells plot also changes. Sometimes it is flat along multiple consecutive trajectory points and sometimes it jumps from present to next trajectory point. Flatness of the plot shows that the number of victim cells will remain same even though spot beam moves forward

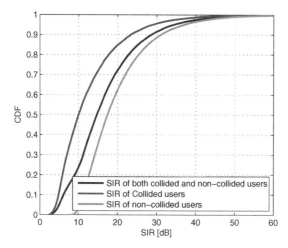

Fig. 5. Received SIR level with and without spot beams drift

and overlapped terrestrial cells. In this case, the frequency of overlapped terrestrial cells is orthogonal to the frequency of the drifted spot beam. On the other hand, increment of number of victim cells will occur when there is co-channel interference between terrestrial cells and spot beam. Figure 4 also shows the impact of influence of single or multiple spot beams on the service area. For example, at 90° drift angle only one spot beam $\{f_1\}$ will be interfere with the co-channel terrestrial cells. On the other hand, when the drift angles are 0° and 45°, two spot beams $\{f_4, f_5\}$ and $\{f_4, f_6\}$ respectively will interfere with the terrestrial cells. Therefore, specially at the higher drift magnitude, number of victim cells increase more when two spot beams overlap the terrestrial cells then by one spot beam.

Fig. 5 shows the three sets of CDF plots of received SIR levels. Green CDF plot shows the SIR level when there is no drift of spot beams. Blue CDF plot shows the collection of SIR level for collided users as well as for non-collided users. In practice, when spot beams drift only co-channel terrestrial cells will be victim and others terrestrial cells will be free from satellite interference even though overlap by drifted spot beams. Finally, red curve CDF plot shows the SIR level only for victim users. It is clearly seen that there is a impact of satellite interference on the performance of integrated satellite-terrestrial system. For example, if the minimum required SIR level is 6.4 dB, then it is seen that with around 24 percent probability the collided users will have less than 6.4 dB. Ultimately the users those who have less than 6.4 dB SIR level in the victim cells will go to outage.

6   Conclusions

In this paper, we have studied downlink interference issues due to pointing errors of spot beam in an integrated satellite terrestrial network. Pointing errors of spot beams characterized by two key parameters: drift magnitude, drift angle. Apart from pointing errors parameters other key parameters associate to model integrated satellite-terrestrial system such as protection ring size, frequency reuse factor of both satellite spot beam and terrestrial cell are also taken into account. Finally simulation has been performed to evaluate the performance of such network under the influence of spot beam by taking into account performance metrics such as victim cells, and SIR. Simulation results helped to understand that there exist significant inter-component interference due to drift of spot beam. Moreover, simulation results reveal that as the drift magnitude of the spot beam increases the performance of integrated satellite-terrestrial system decreases. Hence, it is worthwhile to further investigate the inter-component interference issues by taking into account more detailed physical layer model, different network load and simulate for various application scenarios.

In summary, we can conclude that as the pointing errors of spot beam are inevitable it is also worthwhile to study and find the possible solutions to mitigate and/or avoid these interferences for future development as well as deployment of integrated satellite-terrestrial system.

References

1. Deslandes, V., Tronc, J., Beylot, A.L.: Analysis of interference issues in Integrated Satellite and Terrestrial Mobile Systems. In: Advanced Satellite Multimedia Systems Conference (ASMA) and the 11th Signal Processing for Space Communications Workshop, pp. 256–261 (September 2010)
2. Daoud, F.: Hybrid satellite/terrestrial networks integration. Computer Netw., 781–797 (2000)
3. Ahn, D. H., Kim, J., Ahn, J., Park, D.C.: Integrated/hybrid satellite and terrestrial networks for satellite IMT-Advanced services. International Journal of Satellite Communications and Networking (2010)
4. Evans, B., et al.: Integration of Satellite and Terrestrial Systems in Future Multimedia Communications. IEEE Wireless Communications magazine (October 2005)
5. Stojce, D.I.: Global Mobile Satellite Communications for Maritime, Land and Aeronautical Applications. Springer (2005)
6. ETSI TR 102 662 V1.1.1 (2010-03): Satellite Earth Stations and Systems (SES); Advanced satellite based scenarios and architectures for beyond 3G systems, http://www.etsi.org
7. Alvaro, V., Krauss, H., Hauck, J.: Empirical Propagation model for WiMax at 3.5 GHz in an urban environment. In: Microwave and Optical Technology Letters, pp. 483–487. IEEE Press (2008)
8. Karabinis, P.D., Dutta, S., Chapman, H.H.: Interference Potential to MSS due to Terrestrial Reuse of Satellite Band Frequencies. In: International Communications Satellite Systems Conference, ICSSC (September 2005)

On the Wavelet Families for OFDM System - Comparisons over AWGN and Rayleigh Channels

Ogbonnaya O. Anoh, Raed A. Abd-Alhameed, Steve M.R. Jones,
Yousef A.S. Dama, and Mohammed S. Binmelha

Mobile and Satellite Communication Research Centre (MSCRC),
University of Bradford, United Kingdom
{o.o.anoh,y.a.s.dama,m.s.binmelha}@student.bradford.ac.uk,
{r.a.a.abd,s.m.r.jones}@bradford.ac.uk

Abstract. In the study of OFDM systems, discrete wavelet transforms have been reported to perform better than Fast Fourier Transform in multicarrier systems (MCS) - in terms of spectral efficiency because they can operate without a cyclic prefix, have reduced side-lobes and improved BER. However all of the wavelet families do not perform alike. This study has investigated various wavelet families such as Daubechies, Symlet, Haar (or db1), biorthogonal, reverse-biorthogonal and Coiflets for OFDM system design over an AWGN and multipath channels. Results show that Daubechies, Symlet, Haar and Coiflet wavelet families perform considerably better than other families considered, thus these families could be better in OFDM.

Keywords: DWT, Family, OFDM.

1   Introduction

While wavelet transforms are seriously being considered for OFDM systems, it is pertinent to heuristically select the best among the numerous available members of the wavelet families. The criteria upon which any selection should be made must offer the best possible trade-off in comparison to its other family members. Beside other prominent applications, OFDM is used in Digital Video Broadcasting, DVB [1], which is of the form DVB-C for cable, DVB-S for satellite and DVB-T for terrestrial communications respectively. In [2], the discrete wavelet transform, DWT, has been applauded for its application advantages in video compression, Internet communication compression, object recognition, numerical analysis and signal processing. DWT does not require a cyclic prefix [3] and due to its longer basis functions and reduced side-lobes, bit error ratio (BER) in wavelet based OFDM is improved [4]. These advantages have been possible since the wavelet transform smartly eludes the limitation of non-simultaneous representation of signals in frequency and time. Recall that the Heisinberg uncertainty principle posits that it is impossible to represent a signal as a single point in time and frequency. However, wavelets use a multiresolution, time-frequency representation of a signal called time-scale (or translation-scaling) representation. To do that, a signal, $s(t)$ can be observed

P. Pillai, R. Shorey, and E. Ferro (Eds.): PSATS 2012, LNICST 52, pp. 61–68, 2013.

using this translation-scaling relation such that the signal is scaled and translated (shifted) periodically [5][6]. However, there are limits to the advantages of the wavelets transform in multicarrier system design. Continuous wavelet transforms, S_{CWT} has been defined as [6][7],

$$S_{CWT}(k,\tau) = \int s(t)\psi^*_{k,\tau}(t)dt \tag{1}$$

where * is a complex conjugate. $\psi_{k,\tau}(t)$ is the mother wavelet from which all other basis functions- daughter wavelets used in transformation are derived, through scaling (dilation or compression) and translation (shifting) [7]. It is given by [6], that

$$\psi_{k,\tau}(t) = \frac{1}{\sqrt{k}}\psi\left(\frac{t-\tau}{k}\right) \tag{2}$$

k is a scaling factor and τ is a translation factor. In its continuous form, the expressions are redundant infinite in number and have no analytical (closed-form) solution. As such, they are unsuitable for practical application in MCSs. We can solve this problem by making the wavelets discrete, i.e. by modifying (2) to be of the form [6];

$$\psi_{j,n}(t) = \frac{1}{\sqrt{k_0^j}}\psi\left(\frac{t-n\tau_0 k_0^{\,j}}{k_0^j}\right) \tag{3}$$

And its basis function given by;

$$\int\psi_{j,n}(t)\psi^*_{w,u}dt = \begin{cases}1 & if\ j=w\ and\ n=u \\ 0 & otherwise\end{cases} \tag{4}$$

Where j, n, w and u are integers and $k_0 > 1$ is a fixed dilation step on which the translation factor, τ_0 depends. Precisely, k_0 is usually set to 2 and τ_0 as 1 to satisfy the dyadic property. [8] explained that the dilation problem can be solved by decomposing both the low-pass and high-pass filtered signals alike.

In the following sections, we introduce the DWT in OFDM. Then, we discuss the simulation environment in which this investigation was carried out. Graphs depicting results from this study are then presented, showing how the different wavelet families of different order perform. Using these comparisons, a selection from the wavelet family was then used to compare FFT-OFDM and DWT-OFDM. In the final section, we conclude the study and a list of references follows.

2   DWT for OFDM Systems

2.1   The DWT Scheme

For use in MCS such as OFDM, DWT is seen [2][6] as consisting of a quadrature mirror filter (QMF) bank with low-h and high-g pass filters that convolve with the signal   according   to   the   following   scheme; $s_{low}[n] = h[n] * s[n]$ and $s_{high}[n] = g[n] * s[n]$, where $s[n]$ represents the observed signal. The low-pass filter is related to the high-pass filter as follows, $h[n] = (-1)^n g[1-n]$. $s[n]$ is analyzed and reconstructed according to the Fig. 1;

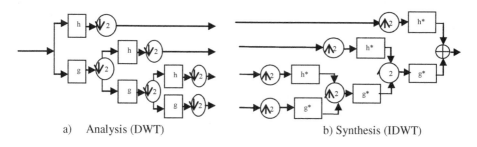

a)   Analysis (DWT)   b) Synthesis (IDWT)

Fig. 1. The DWT/IDWT Processes

In Fig. 1a, the signal is decomposed, filtered and then downsampled. The reverse of this process is done in the receiver wherein the resultant parallel QAM or PSK-modulated samples are first upsampled and then filtered accordingly. This is called reconstruction. In OFDM, however, IDWT and DWT are recursively applied to the parallely aligned signals as long as decomposition and reconstruction length (the leaves of the tree and level) of the signal permits, as in the following block diagram;

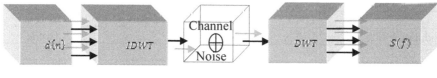

Fig. 2. DWT-OFDM

The randomly generated data are recursively up-sampled and filtered by g* and h* (reconstructed) respectively and passed through the channel with additive white Guassian noise (AWGN). The resultant signal is then decomposed and down-sampled (DWT) and then recovered according to the designer's mapping scheme of choice.

2.2   Decomposition and Reconstruction Conditions

Fig.1 shows a 3-level schematic decomposition and reconstruction structure of a typical DWT for multicarrier systems. Decomposed wavelets can be reconstructed [9] if the energy of the wavelets satisfies this property;

$$\alpha \| s[n] \|^2 \leq \sum_{j,k} | \langle s[n], \psi_{j,k} \rangle |^2 \leq \beta \| s[n] \|^2 \qquad (5)$$

Where α and β are the lower and upper energy limits independent of $s[n]$. Eq. 5 suggests that the signal energy $\| s[n] \|^2$ must be in the positive bound but not infinite which is the basic characteristic of natural signals. The signal, $s[n]$ under study can be perfectly reconstructed according to Fig. 1b and the following [6];

$$h(z)h^*(z^{-1}) + g(z)g^*(z^{-1}) = 2 \qquad (6a)$$

$$h(z)h^*(-z^{-1}) + g(z)g^*(-z^{-1}) = 0 \qquad (6b)$$

h(z) and $g(z)$ are the lowpass and highpass synthesis filters respectively, h*(z^{-1}) and $g^*(z^{-1})$ are the analysis lowpass and highpass filters respectively. With Fig. 1, sub-carrier frequency increase with increase in the decomposition level, thus there would be more resolution and low frequency components.

3   Simulation Environment and Results

To succinctly distinguish among the DWT member families, BPSK has been used to modulate the data transmitted over Rayleigh fading channel with AWGN. Results below show the performance of different discrete wavelet families when compared for BER and SNR. Meanwhile, the following parameters have been used;

Table 1. Simulation Parameters

	DWT	FFT
Modulation Scheme	BPSK	BPSK
FFT Size	Nil	64
Cyclic Prefix	Nil	25%
DWT Families	db1, db2, db3, db5, db44, haar, coif1, coif2, coif3, coif4, sym2, sym3, sym4, sym5, bior1.1, bior1.3, bior1.5, bior2.2, rbio1.1, rbio1.3, rbio1.5, rbio2.2	Nil
Decomposition level	$k = log2\ (N)$, N = 64.	Nil
Symbol length	2*10^4	2*10^4

In DWT, transmitted data must be in order of $M = 2^k$ to ensure possible decomposition and reconstruction. This has influenced our choice of $N = 64$.

3.1   AWGN Channel Only

In FFT-OFDM, if the bandwidth length, N is extended by L-cyclic prefix, it costs an effective bandwidth of $N/(L + N)$ and signal power proportionate to $N/(L + N)$

[10]. We compensate for this power cost with an equivalence of $sqrt(N/L + N)$ and append same to the transmitted signal. This has been cared for in the simulation with Rayleigh channel where cyclic prefix is required for channel immunity. This cost and scaling is not necessary in the DWT-OFDM since the scheme requires no cyclic prefix. The graphs below depict the DWT families and FFT-OFDM compared over an AWGN channel;

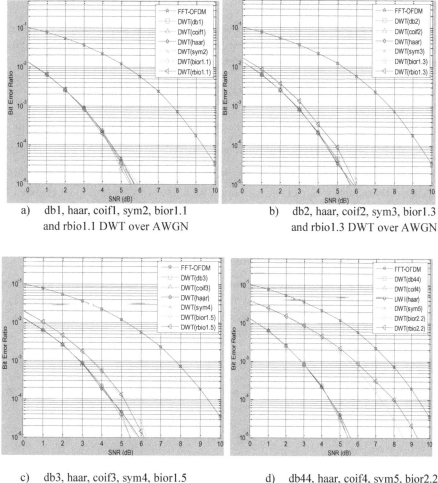

a) db1, haar, coif1, sym2, bior1.1   b) db2, haar, coif2, sym3, bior1.3
    and rbio1.1 DWT over AWGN   and rbio1.3 DWT over AWGN

c) db3, haar, coif3, sym4, bior1.5   d) db44, haar, coif4, sym5, bior2.2
    and rbio1.5 DWT over AWGN   and rbio2.2 DWT over AWGN

Fig. 3. Performance of DWT families and FFT-OFDM over AWGN channel

Using the enumerated parameters, the signal is reconstructed for $log2(N)$ or 4-levels in our case, with $log2(N) + 1$ leaves of the wavelet trees. In the receiver, the received signal is again decomposed up to 4-levels with $log2(N) + 1$ leaves. These leaves form the subchannels of the MCS and these subchannels modulate the data over the transmitter and the receiver. Over AWGN channel, results show that different wavelet families perform differently with Daubechies family performing best, while the biorthogonal and reverse-biorthogonal families perform worst. Since different families posses different filters, thus different wavelet families must perform differently. However, other family members, such as Haar, Coiflet, and Symlet perform likely equally with the Daubechies family members.

3.2   Rayleigh Fading Channel

In [11], a Rayleigh fading channel is shown to consist of several multipath channels. It is characterized by a number of attenuated, time delayed copies of the original transmitted signal. In the baseband, its impulse response can be modeled as [12];

$$h(t,\tau) = \sum_{n=1}^{N} a_n(t,\tau)e^{-j\theta_n} \delta(t - \tau_n(t)) \tag{6}$$

$a_n(t,\tau)$ and τ_n are the amplitude and time delay respectively with a phase shift of $\theta_n = 2\pi f_c \tau_n(t)$ for the nth multipath at a prevailing time, t. N is the possible number of multipath and $\delta(\bullet)$ is the Dirac delta. Assuming no line of sight [13][14], Eq.6 correctly models a Rayleigh distribution for a time varying multipath channel. In the frequency domain, the channel transfer function, H, is obtained as a Fourier transform pair of the channel impulse response.

3.3   The Channel Estimation Scheme

In [7], Zero-forcing, ZF channel estimation method in time domain of wavelet-OFDM for feedforward systems has been shown for In-home Power Line Communication, PLC channel. We extend this ZF channel estimation in time domain of a DWT-OFDM system to a Rayleigh fading channel with AWGN noise.

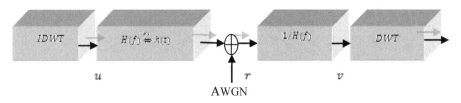

Fig. 4. The ZF channel estimation scheme

Given that r is the received signal, expressed as; $r = Hs + w$, where w is the AWGN. If we define the channel impulse response of the systems as h, its transfer function is a Fourier transform of h and is given by H. Since the DWT offers a long basis function thus promising a total ISI and ICI combat, no CP is thus required so that we equalize the channel just before the DWT operation into frequency domain according to the following; $v = r.(1/H)$. By the above scheme, the result below shows the performance of db5, bior2.2 and FFT-OFDM in a Rayleigh fading channel.

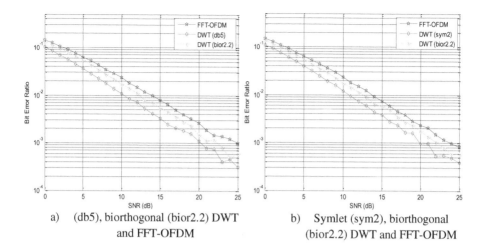

a) (db5), biorthogonal (bior2.2) DWT   b)   Symlet (sym2), biorthogonal
and FFT-OFDM   (bior2.2) DWT and FFT-OFDM

Fig. 5. Performance comparison of Daubechies (db5), Symlet (sym2), biorthogonal (bior2.2) DWT and FFT-OFDM over Rayleigh fading channel with AWGN

Comparing the db5 and sym2 with the FFT for instance, there is nearly 5.0dB gain (SNR) and with the bior2.2, there is up to 3.0dB gain (SNR). Thus, we see that the db5 outperforms all the family members considered in this respect. It then follows that Daubechies DWT family members perform better than other DWT family members.

Conclusion

An in-depth study into the performance behaviour of discrete wavelet transform for multicarrier systems, such as OFDM, has been presented. Results show that DWT of the Daubechies families champion in performance when observed in terms of BER and SNR. Haar DWT which is on the other the db1 sub-family can be considered alongside the Daubechies families. Except for biorthogonal and the reverse birothogonal, all other wavelet families studied perform alike. Within the Daubechies DWT families, lower order sub-families which have lower number of filters can be considered for OFDM systems in terms of improved BER and run-time at the expense of the higher order members.

References

[1] Van Nee, R., Prasad, R.: OFDM for Wireless Multimedia Communications. Artech House, London (2000)
[2] Weeks, M.: Digital Signal Processing Using MATLAB and Wavelets, 2nd edn. Jones and Bartlett Publishers, Sudbury Massachusetts (2011)

[3] Sharma, S., Kumar, S.: BER Performance Evaluation of FFT-OFDM and DWT-OFDM. International Journal of Network and Mobile Technologies 2(2) (2011)

[4] Khan, U., Baig, S., Mughal, J.: Performance Comparison of Wavelet Packet Modulation and OFDM for Multipath Wireless Channel with Narrowband Interference. In: 2nd International Conference on Computer, Control and Communication, pp. 1–4 (2009)

[5] Mallat, S.G.: A Theory for Multiresolution Signal Decomposition: The Wavelet Representation. IEEE Transaction on Pattern Analysis and Machine Intelligence 11(17) (1989)

[6] Valens, C.: A Really Friendly Guide to Wavelets;
     `http://polyvalens.pagesperso-orange.fr/clemens/wavelets/wavelets.html`, (November 4, 2011)

[7] Sripath, P.: Efficient Implementations of Discrete Wavelet Transformations using FPGAs, MSc Thesis, Florida State University College of Engineering (2003)

[8] Farrukh, F., et al.: Performance Comparison of DFT-OFDM and Wavelet-OFDM with Zero-Forcing Equalizer for FIR Channel Equalization. In: International Conference on Elec. Engineering, pp. 1–5 (2007)

[9] Daubechies, I.: Ten Lectures on Wavelets, 2nd edn., Philadelphia. CBMS-NSF-Regional Conference Series on Applied Mathematics (1992)

[10] Goldsmith, A.: Wireless Communication. Cambridge University Press (2005)

[11] Rappaport, T.S.: Wireless Communications Principles and Practice, 2nd edn. Prentice Hall, Inc., Upper Saddle River (2002)

[12] Tse, D., Viswanath, P.: Fundamentals of Wireless Communication. Cambridge University Press (2005)

[13] Parsons, J.D.: The Mobile Radio Propagation Channel, 2nd edn. John Wiley and Sons Limited, England (2000)

[14] Yang, H.: A Road to Future Broadband Wireless Access: MIMO-OFDM-Based Air Interface. IEEE Communications Magazine (2005)

Performance Analysis and Optimization of Downlink Multi-User MIMO LTE for Satellite Communications

Constantinos T. Angelis[1,*] and Spiros Louvros[2]

[1] Department of Informatics and Telecommunications
Technological Educational Institute of Epirus, Arta, Greece
kangelis@teiep.gr
[2] Department of Telecommunication Systems and Networks
Technological Educational Institute of Messologi, Messologi, Greece
slouvros@teimes.gr

Abstract. This article presents a theoretical analysis for packet delay and simulation results for a realistic implementation of the Multi-User MIMO LTE Release 8 downlink standard in mobile satellite systems. Two Satellites, one Ground Station and two 2x2 MIMO Channels have been used as a basic configuration in the simulation scenarios and various key characteristics of the MIMO channel and the LTE radio interface, including physical layer and radio resource management functions ware simulated and their impact on system performance is evaluated for moving terminals in wide area scenarios. Simulation results suggest that in practice multi-user LTE, when applied to the transmission over satellite links, is able to support multi stream transmission with very high data rates, even for hand held moving terminals. Moreover, the improvements of Multi-User MIMO transmissions for different system configurations are clearly shown for different number of users. Finally a theoretical approach, considering OFDMA scheduler functionality, is presented leading into an optimization procedure for MAC transport channel packet length.

Keywords: LTE, Satellite, Multi-User MIMO, OFDMA, MAC transport channel.

1 Introduction

The growing demands for broadband wireless data communications, in multihop capable interfaces, are becoming more and more intense due to their improvements of coverage and capacity. For these reasons they are proposed for the next generation cellular systems like 3G-LTE [1]. This has motivated many research efforts in the last years, puts high pressure on operators to increase the capacities of their networks and on the industry for enabling such an increase also in the long term future via more efficient and flexible communication standards. Long-Term Evolution (LTE) is an emerging radio access network technology standardized in 3GPP [1], that meets all the previous constrains, and evolving as an evolution of Universal Mobile

---

* Member ICST, IEEE, EuMA.

P. Pillai, R. Shorey, and E. Ferro (Eds.): PSATS 2012, LNICST 52, pp. 69–80, 2013.

Telecommunications System (UMTS). LTE uses Orthogonal Frequency Division Multiplexing OFDM as downlink air interface multiple access scheme [1].

The interesting research outcomes that obtained by proposed LTE techniques in terrestrial networks have engender further interest in investigating the possibility of applying the same principles in satellite networks. However, there are fundamental differences between satellite and terrestrial channels that lead this effort to a very challenging research subject. The most important restriction for the successful deployment of mobile satellite systems in the LTE networks is the overall cost minimization. In order to achieve this, the technological commonalities with the terrestrial networks have to be maximized. This can be done by considering the terrestrial radio interface as the baseline for the satellite radio interface, introducing only those modifications that are strictly needed to deal with the satellite peculiarities. The most important peculiarities are nonlinear distortion introduced by the on-board power amplifiers, long round-trip propagation times, and reduced time diversity.

Many researchers work on LTE and MIMO based systems for satellite communications. General LTE concept descriptions are available in [2-7]. In these papers, the focus is on key characteristics of the LTE radio interface. A set of such key characteristics are both qualitatively discussed and quantitatively evaluated in terms of downlink user data rates, spectrum efficiency generated by means of system level simulations and measurements. In [2] the applicability of the 3GPP Long Term Evolution standard to mobile satellite systems is investigated. In [5] the applicability of multiple-input multiple-output (MIMO) technology to satellite communications at the Ku-band and above is investigated. In [6] a three dimensional channel model for the study of distributed MIMO communication systems is presented. Finally, in [7] a novel physical-statistical generative model for the land mobile satellite (LMS), dual polarized, MIMO channel along tree sided roads is presented.

Performance optimization of such a satellite network is crucial due to long propagation delays, contributing into throughput reduction. MAC layer maps the logical channels into a new channel format called transport channel [9]. MAC packet length is dynamically decided based on the service and the used Modulation and Coding scheme (MCS). Optimizing the packet length will contribute into less delays thus improving the overall performance. In [10] combination of power and time-frequency domain allocation of resources has been extensively studied and in [11] different scheduling time-frequency domain algorithm approaches have been also proposed. Finally in [12] a good algorithmic analysis has been presented exploiting simultaneously throughput optimization with QoS rate restrictions. Most important scheduling algorithms involved in existing analysis in literature are considering mostly proportional fair schemes, max CQI, min interference or round robin and performance results have been proposed based on the channel conditions. In order to make good scheduling decisions, uplink/downlink scheduler should be aware of channel quality in the time-frequency domain, as channel fading is time-frequency dependant, using channel quality reports (CQI) [13]. Although ideally, the scheduler should have exact knowledge of the channel quality for each sub carrier and each user [14], this is not realistic due to increased signalling load and a compromise between good channel quality knowledge and CQI reports has to be decided [15]. However all previous analysis mostly focus on the scheduling algorithm performance optimization based on algorithm simulations and simple theoretical approaches for CQI, quality and MCS selection. What is still missing is an exact analytical model to combine all different aspects of scheduler functionality in order to provide an exact optimization solution for MAC transport channel length.

In this paper a realistic implementation of the Multi-User MIMO LTE Release 8 downlink standard in mobile satellite systems is investigated. Simulation results show that the performance improvement depends on the considered parameters. MAC packet length optimization analysis focuses on basic delay calculation over the LTE air interface based on scheduler and MAC layer functionality. In the basic analysis retransmissions are considered as a general drawback on the throughput optimization. The rest of the paper is organized as follows: Section 2 describes the FDD MIMO LTE system level model and an overview about the simulation model used for the investigation with description of the performed link and system level simulations; Section 3 presents the basic theoretical delay analysis; Section 4 presents the simulation results; and finally the conclusions are given in Section 5.

2   The Multi-User FDD MIMO LTE Simulation Model

This section presents the proposed Multi-User FDD MIMO LTE system level model and the simulation setup environment. The simulation results were obtained with the Agilent Advanced Design System (ADS) [8]. We performed Multi-User TDD downlink coded measurements on MIMO channels for all the combinations of the parameters discussed bellow. In our analysis we assume a scenario in which two GEO satellites broadcast services to a Ground Mobile Terminal (GMT). Both satellite and terrestrial components work at the same frequency, thus realizing a Single Frequency Network (SFN). Figure 1 shows a schematic diagram of the simulation model that consists in three main blocks for each satellite link: transmitter, channel and receiver chains.

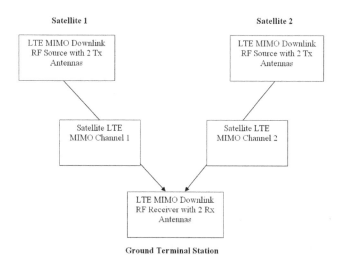

Fig. 1. The Multi-User TDD MIMO LTE simulation model

The reference channel in A.3 of TS 36.101 [1] is used as a signal source with the exception that no Physical Hybrid ARQ Indicator Channel (HARQ) transmissions are employed and a Bandwidth of 10 MHz is employed. The modulation type that we used in our simulation scenarios is the 10 MHz QPSK 1/3. Four (2 per satellite) transmitter antennas and 2 receiver antennas were used as parts of the simulation in Transmit diversity transmission mode. The correlation matrix support was set to Medium. Table 1 shows some of the most fundamental simulation conditions and parameters.

Table 1. LTE System fundamental simulation conditions and parameters

Parameter	Value
Carrier Frequency	2.5 GHz
Bandwidth	10 MHz
Frame Mode	TDD Configuration
Oversampling	Ratio 2
Cyclic Prefix	Normal
Antenna Configuration	2x2 for each Satellite
Number of code word(s)	2 (Spatial multiplexing)
Number of layer(s)	2 (Spatial Multiplexing)
Correlation	Medium ($\alpha=0.3$, $\beta=0.9$)

Table 2. UEs and GMT fundamental simulation conditions and parameters

Parameter	Case 1 1 User per Satellite	Case 2 2 Users per Satellite	Case 3 3 Users per Satellite
Velocity	**(a)** 5km/h, **(b)** 50km/h	**(a)** 5km/h, **(b)** 50km/h	**(a)** 5km/h, **(b)** 50km/h
Number of Subframes	10	10	10
Modulation	QPSK	QPSK	QPSK
UEs Payload	1/3	1/3	1/3
Preconding	Yes	Yes	Yes
Channel Estimator	2D MMSE	2D MMSE	2D MMSE
Number of RBs for each 2D-MMSE interpolation	5	5	5
Turbo decoder iteration	6	6	6

The basic configuration of the MIMO channels is in accordance to the specifications of Annex B - TS 36.101 [1]. The MIMO satellite channel was modeled as a normal MIMO LTE channel with medium correlation and the Doppler frequency was taken into account. In addition, to accurately model the MIMO satellite channel, a High Power Amplifier (HPA) is introduced within each satellite channel [2-4]. The HPAs are assumed to work with an Input Back-Off (IBO) of 4dB. Both Satellite channels assumed to be identical for simplicity. Moreover the proposed inter-TTI interleaving technique that was proposed in [2] is used in order simulate the satellite propagation conditions.

In the proposed scenario the Ground Mobile Terminal (GMT) motion is taken into account by considering the Doppler spread. Two different speed values are analyzed, that are $v = 5$km/h and $v = 50$km/h. The GMT receiver uses a 2D MMSE channel estimator with 5 RBs for each 2D-MMSE interpolation and the Turbo decoder iteration number was set to 6. Simulations were performed for both Single and Multi User UEs. The parameters used in the simulations are given in Table 2.

3   Basic MAC Delay T_{delay} Analysis

In LTE services traffic is based on IP technology [1]. Between UE and Nodeb (terrestrial or satellite) each MAC packet is supposed to be transmitted completely over the air interface before starting transmission of next MAC packet in a duration of Time Transmission Interval TTI=1ms. In such a case the advantage is that LTE could provide error detection and retransmission not only on TCP/UDP IP level on the core network but also on the air interface resulting into better correction techniques improving thus the throughput and minimizing delays [1]. In each downlink TCP/UDP IP initiation session, scheduling decisions are mostly decided based on QoS service profile and Channel Quality Index (CQI) measurements. A number of variable length MAC packets arriving at the destination will be acknowledged by sending back a short acknowledgment packet of length M_{ack} bits size on the PDSCH downlink channel. There is supposed to be a window size of length W, meaning that on the eNodeB receiver an acknowledgment is created only when a total number of W MAC packets are received.   One interesting question that arises then is which is the optimum MAC packet length size in order to improve throughput per connection. Suppose that a TCP/UDP IP packet of M_I bits per packet be framed in such a way that resulting MAC packets of variable length M_{mac} bits per packet contains a fixed number of M_{over} header bits per packet [16]. In such a model then one M_I packet will be segmented into Int[M_I / M_{mac}] MAC packets. Definitely the division will never allocate an integer number and hence padding should be applied in order to be able to fix the leftover bits into exactly Mmac packet size. The total number of MAC packets out of one M_I size TCP/UDP IP packet are Int[M_I / M_{mac}] +1, the total number of MAC bits to be transmitted out of one M_I size TCP/UDP IP packet are M_I + {Int[M_I/ M_{mac}] +1}M_{over}, where the factor {Int[M_I / M_{mac}] +1}M_{over} is the overhead created by MAC layer for the M_I size TCP/UDP IP packet transmission. What is then interesting to measure is the total transmission time (as a measure of expected delay) and the error performance as functions of MAC packet size M_{mac} and to find the best compromise. Expected whole TCP/UDP IP packet transmission time in ideal conditions, without retransmissions, is:

$$
T_{delay} = \frac{\left\{ M_I + \left[Int\left(\dfrac{M_I}{M_{mac}} \right) + 1 \right] M_{over} + F \right\}}{M \cdot N \cdot r_{TTI}} T_s + \langle n \rangle T_s \qquad (1)
$$

where $M_I + \left[Int\left(\dfrac{M_I}{M_{mac}} \right) + 1 \right] M_{over} + F$ is the total number of TCP/IP bits to be

transmitted considering also overhead M_{over} and padding bits,

$F = M_{mac} + \left(int\left[\dfrac{M_I}{M_{mac}} \right] + 1 \right) \cdot M_{mac} - M_I$ considering always that M_{mac} packet size

is always less than the M_I IP packet size. Also $\left[Int\left(\dfrac{M_I}{M_{mac}W} \right) + 1 \right] A$ is the number of

bits to be transmitted on the downlink PHICH physical channel per W window size

which is not delayed by scheduling decisions ($\langle t_{sch}^{ack} \rangle = T_s = 1$ ms). r_{TTI} is the number

of transmitted bits per scheduled block (TTI = 1ms and bandwidth of 180 kHz) which
depends on Link Adaptation Modulation Scheme, N is the average allocated number
of 180 kHz radio block (RB) units of bandwidth per TTI considering also the
constraint that $0 \leq N \leq BW$ where BW is the allocated bandwidth in the cell
planning (minimum 1.8 MHz up to maximum 20 MHz) and M is the number of
antenna ports (in case of MIMO implementation). In this analysis then we consider,
for average $N \cdot 180kHz$ allocated RB, $N \cdot r_{TTI}$ bits out of total o be transmitted
simultaneously in TTI = 1 ms and finally if we have spatial multiplexing of $M \times M$
MIMO antenna ports then $M \cdot N \cdot r_{TTI}$ bits are expected to be transmitted
simultaneously in TTI = 1 ms. Finally $\langle n \rangle \cdot T_s$ is the average schedule time by
scheduler, where n is an integer value to indicate the number of subframes (Time
Transmission Interval $T_s = 1ms$) that one MAC packet is not scheduled by scheduler
in a total scheduling period T. Remember that n for downlink scheduling decisions
depends mainly on the QoS Guaranteed Bit Rate (average) GBR parameter, on
Channel Quality Index CQI measurement report and also on UE transmitter mean
packet waiting time on the buffer. Substituting F into equation (1) we get:

$$
T_{delay} = \frac{\left\{ \left[Int\left(\dfrac{M_I}{M_{mac}} \right) + 1 \right] M_{over} + M_{mac} + \left(int\left[\dfrac{M_I}{M_{mac}} \right] + 1 \right) \cdot M_{mac} \right\}}{M \cdot N \cdot r_{TTI}} T_s + \langle n \rangle T_s \qquad (2)
$$

Considering then that its MAC packet could be retransmitted maximum v times and
the average transmission rate is m_{mac} from (1) TCP/IP packet transmission delay
time could be recalculated as:

$$T_{delay}^{retr} = \frac{\left\{ m_{mac} \cdot M_I + m_{mac} \cdot \left[Int\left(\frac{M_I}{M_{mac}} \right) +1 \right] M_{over} + F \right\}}{M \cdot N \cdot r_{TTI}} T_s + \langle n \rangle T_s \qquad (3)$$

Substituting then F padding bits from previous analysis TCP/IP packet transmission delay time could be recalculated as:

$$T_{delay}^{retr} = \frac{\left\{ M_I (m_{mac} -1) + m_{mac} \cdot \left[Int\left(\frac{M_I}{M_{mac}} \right) +1 \right] M_{over} + M_{mac} + \left(int\left[\frac{M_I}{M_{mac}} \right] +1 \right) \cdot M_{mac} \right\}}{M \cdot N \cdot r_{TTI}} T_s + \langle n \rangle T_s \quad (4)$$

Average number of retransmissions m_{mac} is a function of the MAC packet error rate. In this analysis we do consider p to be the MAC packet successful acceptance probability rate. If after maximum v transmissions the MAC packet is still corrupted it will be finally forwarded to the upper RLC layer with probability $v \cdot (1-p)^v$ since MAC layer does not discard corrupted packets even after the maximum v number of retransmissions. Hence overall average retransmission rate is calculated as:

$$m_{mac} = v(1-p)^v + \sum_{k=1}^{v} kp(1-p)^{k-1} = \frac{1-(1-p)^v}{p} \qquad (5)$$

Since $(1-p) = 1 - (1 - p_b)^{M_{mac}}$ where p_b is the bit error rate which could be substituted with BER and should be calculated from simulations (as we did on section 4 in Fig. 6a) or real channel measurements.

$$m_{mac} = \frac{1-(1-p)^v}{p} = \frac{1-\left(1-(1-p_b)^{M_{mac}}\right)^v}{\left(1-p_b\right)^{M_{mac}}} \qquad (6)$$

Retransmission rate m_{mac} depends on number of attempts v and on the size of the MAC packet M_{mac}. LTE MAC Scheduler follows rules of priorities on scheduled packets per TTI. Basic priority rule assigns retransmission packets with highest priority rather than new packets on the transmitter bucket when they are scheduled by scheduler. Expected delay budget equals:

$$\max \tau_{delay}^{budget} = v_{max} T_s + \langle n \rangle T_s \Rightarrow v_{max} = \frac{\max \tau_{delay}^{budget} - \langle n \rangle T_s}{T_s} \qquad (7)$$

and equation (6) finally becomes:

$$m_{mac} = \frac{1-\left(1-(1-BER)^{M_{mac}}\right)^{\frac{\max \tau_{delay}^{budget} - \langle n \rangle T_s}{T_s}}}{\left(1-BER\right)^{M_{mac}}} \qquad (8)$$

4   Results and Discussion

Performance of the simulated schemes is compared in terms of Tx and Rx Signal Spectrum, Complementary Cumulative Distribution Function (CCDF), BER and BLER measurements. Simulations were conducted taking into consideration the number of users in each Satellite Link. Figure 2 presents indicative Signal spectrums in the 4 transmit antennas (2 per Satellite) and in the 2 receive antennas for $E_b / N_o = 20dB$. The modulation scheme in figure 2 is QPSK 1/3, while the correlation was set to medium. Figures 3 and 4 present CCDF measurements of the TDD Downlink LTE signals in the 2 receive (Rx) antennas. The modulation scheme is QPSK 1/3. As it is seen for figure 3 the number of EUs increment does not affect the CCDF measurements for both ground terminal velocities of 5km/h and 50 km/h. On the other hand the ground terminal velocity affects the CCDF measurements as it is seen in figure 4.

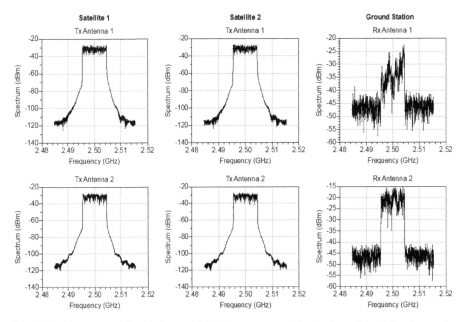

Fig. 2. Signal spectrums in the 4 transmit (Tx) antennas and in the 2 receive (Rx) antennas for $E_b / N_o = 20dB$

BER, BLER and Throughput measurements where performed for the case of QPSK 1/3 modulation scheme. The throughput was evaluated using the proposed specifications of the LTE radio interface, where the modulation was fixed at QPSK. The turbo code that was used and the mapping from code words to layers ware according to LTE specifications. The throughput was calculated as the product:

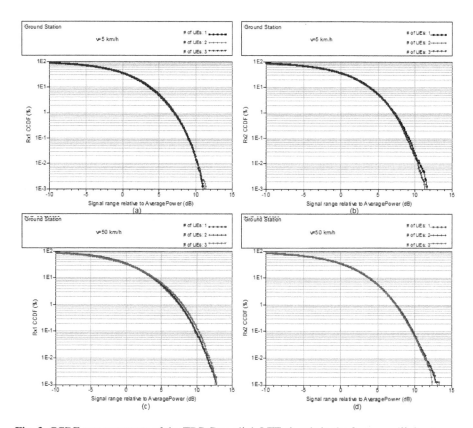

Fig. 3. CCDF measurements of the TDD Downlink LTE signals in the 2 receive (Rx) antennas. (a) and (b) stands for v=5km/h, (c) and (d) stands for 50km/h

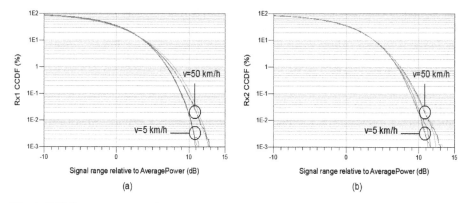

Fig. 4. CCDF measurements of the TDD Downlink LTE signals in the 2 receive (Rx) antennas

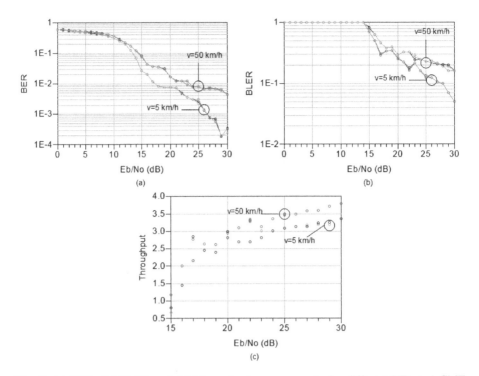

Fig. 5. (a) BER, (b) BLER and (c) Throughput measurements for different UEs and GMT velocities

$$Throughput = (1 - FER) * N_{layers} * CodingRate \qquad (9)$$

where FER denotes the frame error rate, N_{layers} denotes the number of layers and CodingRate is the rate of the Turbo code. Figure 5 shows BER, BLER and Throughput measurements of the TDD Downlink LTE signals in the first Satellite link. As it is seen BER and BLER measurements are degraded when the velocity of the GMT increases, while when the number of UEs increases the BER and BLER values remain unchanged. On the other hand Throughput practically remains unchanged. These results indicate that the increment of the multi-users number, from one to tree, does not have any impact in the overall performance for the given configuration. The CCDF, BER, BLER and Throughput simulation results that derived from figures 3 to 5 indicate that the proposed Multi-User MIMO LTE downlink can handle up to 3 UEs per Satellite link.

Substituting results of BER into equation (9), for a specific service maximum allowed delay $\max \tau_{delay}^{budget}$, for specific channel conditions (thus scheduler average scheduling instances $\langle n \rangle$ and for $T_s = 1ms$ as the sub-frame time transmission interval we could calculate the average number of retransmissions m_{MAC}. Then substituting into equation (4) we could calculate the optimum M_{mac} length to keep delay T_{delay} into the desired bounds.

5   Conclusion

In this paper system performance of a Multi-User MIMO LTE downlink for Satellite Communications which consists of Two Satellites, one Ground Station and two 2x2 MIMO Channels was investigated and various key characteristics of the MIMO channel and the LTE radio interface, including physical layer and radio resource management functions ware simulated and their impact on system delay and performance metrices is evaluated for moving terminals in wide area scenarios. Simulation results indicate that the proposed Multi-User LTE system, when applied to the transmission over satellite links, is capable of supporting multi stream transmission with very high data rates, even for hand held moving terminals for up to 3 UEs per Satellite link.

Acknowledgments. The Author would like to thank Agilent, Inc. for the license to use ADS for educational and research purposes.

References

1. TS 36.101 3rd Generation Partnership Project; Technical Specification Group Radio Access Network; Evolved Universal Terrestrial Radio Access (E-UTRA); User Equipment (UE) radio transmission and reception (Release 8)
2. Papalco, M., Neri, M., Vanelli-Coralli, A., Corazza, G.E.: Using LTE in 4G Satellite Communications: Increasing Time Diversity through Forced Retransmission. In: 10th International Workshop on Signal Processing for Space Communications, Rhodes Island, Greece, October 6-8 (2008)
3. Neri, M., Salmi, P., Corazza, G.E.: Fractional Predistortion Techniques with Robust Modulation Schemes for Fixed and Mobile Broadcasting. In: 13th IST Mobile and Wireless Communications Summit, IST 2004 (2004)
4. Berdondini, M., Neri, M., Corazza, G.E.: Adaptive Fractional Predistortion Techniques for Satellite Systems based on Neural Networks and Tables. In: Proceedings of IEEE 65th Vehicular Technology Conference, VTC SPRING 2007, pp. 1400–1404 (2007)
5. Liolis, K.P., Panagopoulos, A.D., Cottis, P.G.: Multi-Satellite MIMO Communications at Ku-Band and Above: Investigations on Spatial Multiplexing for Capacity Improvement and Selection Diversity for Interference Mitigation. EURASIP Journal on Wireless Communications and Networking, Article ID 59608, 11 pages (2007)
6. Li, X., Liu, Y.: A 3-D Channel Model for Distributed MIMO Satellite Systems. In: IEEE Global Communications Conference (Globecom 2010), Miami, Florida, USA, December 6-10 (2010)
7. Cheffena, M., Perez-Fontan, F., Lacoste, F., Corbel, E., Mametsa, H.-J.: Land Mobile Satellite Dual Polarized MIMO Channel along Roadside Trees, Roma, Italy, April 11-15 (2011)
8. Agilent Advanced Design System (ADS), http://www.agilent.com

9. 3GPP TS 25.104 'Base Station (BS) radio transmission and reception (FDD)'
10. Liu, H., Li, G.: OFDM-Based Broadband Wireless Networks, Design and Optimization. John Wiley & Sons (2005)
11. Monghal, G., Pedersen, K.I., Kovacs, I.Z., Mogensen, P.E.: QoS Oriented Time and Frequency Domain Packet Schdulers for the UTRAN Long Term Evolution. In: Proc. of IEEE Vehicular Technology Conference, Spring, pp. 2532–2536 (2008)
12. Wunder, G., Zhou, C., Bakker, H.-E., Kaminsk, S.: Thoughput Maximization under Rate Requirements for the OFDMA Downlink Channel with Limited Feedback. Article ID 437921, EURASIP Journal on Wireless Communications and Networking (2008)
13. Homayounfar, K., Rohani, B.: CQI Measurement and Reporting in LTE: A New Framework, IEICE Tech. Rep. 108(445), RCS2008-244, 191–196 (2009)
14. Kawser, M.T., Hamid, N.I.B., Hasan, M.N., Alam, M.S., Rahman, M.M.: Downlink SNR to CQI Mapping for Different Multiple Antenna Techniques in LTE. In: International Conference on Future Information Technology (ICFIT), Changsha, China (December 2010)
15. Kolehmainen, N.: Channel Quality Indication Reporting Schemes for UTRAN Long Term Evolution Downlink. In: Proc. of IEEE Vehicular Technology Conference, Spring, pp. 2522–2526 (2008)
16. 3GPP TS 36.321 v8.1.0

Two Miniaturized Printed Dual-Band Spiral Antenna Designs for Satellite Communication Systems

Mohammed S. Binmelha, Chan H. See, Raed A. Abd-Alhameed,
M.S. Alkambashi Alkambashi, D. Zhou, Steve M.R. Jones, and P.S. Excell

Mobile Satellite Communications Research Centre, University of Bradford, Richmond Road,
Bradford, West Yorkshire, BD7 1DP, UK
{ M.S.Bin-melha.chsee2.raaabd}@bradford.ac.uk

Abstract. Two novel reduced-size, printed spiral antennas are proposed for use in personal communications mobile terminals exploiting the "big low earth orbit" (Big-LEO) satellite system (uplink 1.61–1.63 GHz; downlink 2.48–2.5 GHz). The two proposed antenna give 3.12—6.25% bandwidth at lower resonant mode of 1600MHz, while at the higher resonant mode of 2450MHz a bandwidth of around 6% is obtained. The experimental and simulated return losses of the proposed antennas show good agreement. The computed and measured gains, and axial ratios are presented, showing that the performance of the proposed two antennas meets typical specifications for the intended applications.

Keywords: printed spiral antennas, bandwidth, return losses, axial ratios.

1   Introduction

Personal satellite communications systems provide global coverage, especially where there are no nearby terrestrial base stations [1,2]. The majority of systems currently in operation use the 'big low earth orbit' (Big-LEO) satellite system, such as 'Globalstar', which was chosen as a system for detailed study [1]. Handsets of this system require broad-beam radiation patterns with low-cross-polarization, circularly-polarized antennas to get acceptable link margins. These terminals use an uplink band at 1.61-1.6214GHz (L-band) and downlink band at 2.4835-2.5GHz (ISM/S-band).

Satellite mobile communications systems have been available for some years. The systems have experienced some commercial problems, particularly due to the unexpectedly rapid growth of terrestrial systems, but they still have a place in the overall range of wireless communication systems.

The size and appearance of the satellite handset quadrifilar helix antennas and their radomes presents a problem of image and convenience for a public used to the low-profile antennas of terrestrial systems [3-6], whilst the design must achieve specific antenna requirements appropriate for satellite communications.

Reducing the size of the antenna is not easy, since it requires us to have more directive gain than the lowest order (dipole) mode. This causes difficulties if its size is

P. Pillai, R. Shorey, and E. Ferro (Eds.): PSATS 2012, LNICST 52, pp. 81–86, 2013.

required to be less than about a half wavelength at the operating frequency, due to what is effectively a 'law of physics' for small antennas, the so-called Wheeler limit [7]. Some success in reducing the size of antennas has been achieved by coiling the wire elements, first into helices and later into spirals [8-13]. Understanding of traditional circular spirals is well advanced, but square designs are likely to fit more conveniently into practical products [8, 9].

Spiral antennas are particularly known for their ability to produce very wide band, almost perfectly circularly-polarized radiation over their full coverage region. As a result of this polarization characteristic and the ability to produce a broad zenith-directed pattern, spiral antennas are popular for use in satellite mobile communication handsets.

In this paper, the achieveable size of the Quadrifilar Square Spiral Antenna (QSSA) discussed in [8- 10] was significantly reduced by a new design of Dual-Arm Square Spiral Antenna (DASSA) on a thin dielectric substrate. This made the new antenna of a size that would be easily mountable on the top of a handheld terminal for use with a low-earth-orbit (LEO) personal satellite communications network. The present program of work has thus initiated a study of a square version of the dual spiral antenna (DASSA) that should satisfy the following design requirements: (i) hemispherical radiation pattern with elevation coverage from the zenith to a nominal minimum elevation angle (typically 10°and 20°); (ii)circular polarization with an axial ratio better than 5dB within the coverage angle; (iii)operational bandwidth to be covered with one antenna operating by itself, either with a single wide bandwidth or, with the assistance of a simple matching circuit, over the two sub-bands of interest; (iv) the size to be minimized by implementing the DASSA over dielectric substrates of high relative permittivity.

The new compact antenna design for handheld satellite mobile communication is investigated and discussed at L band (1.61-1.6214GHz), ISM S-band (2.4835-2.5GHz) and dual L-S bands. Two different antenna types are presented using two-arm spirals connected at the centre by a small rectangular patch and fed via a stripline from each end. Various stripline widths are studied. The inputs return loss and field patterns show quite reasonable results that satisfy the requirements of the communication strategy. The results in terms of the antenna size and radiation performance are addressed and compared to previous published data.

2   Antenna Design Concept and Geometry

The DASSA is an electrically small antenna providing circular polarization over a broad angular region. The antenna consists of two spirals equally spaced circumferentially (placed at 180° to each other) and fed by equal amplitude signals with 180°-phase difference between feeding sources. The DASSA can also be described as two orthogonal bifilar helix antennas fed in phase quadrature, (where a bifilar is a two-element helix antenna). The two spirals are fed at their ends, so that the feed lines in this case do not cause significant problems from the point of view of mutual coupling effects.

The desirable size for the DASSA will be that of the top of a typical personal handset; however, the initial design was made somewhat larger in order to prove the concept [8]: the work presented here uses a solid dielectric beneath the spirals to reduce the antenna size. All antennas were mounted on a thin dielectric substrate of ε_r = 2.55, tan δ = 0.0018 and thickness of 1.524 mm. Fig.1 shows the geometry of the two proposed antennas for dual-band (L and ISM/S-band) operation. As can be seen, the two antenna sizes and the striplines width used for L and ISM/S bands are (16 × 12 mm^2 and 0.25mm) and (12.5 × 8.4mm^2 and 0.75mm) respectively. These two designs will give a variety of choices for antenna designer to further investigate the required antenna performance. Moreover, from the antenna sizes presented, the antennas can easily be mounted on top of a small area of the handset.

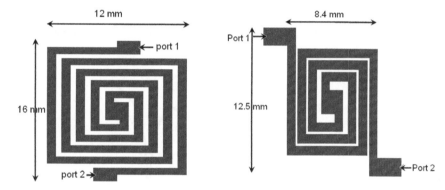

Fig. 1. The geometry of proposed dual-band printed antennas. (Left: Ant1 with stripline width 0.25 mm, Right: Ant2 with stripline width 0.75 mm.)

3   Results and Discussion

The simulated results of all antenna geometries shown in Fig. 1 were carried out using Agilent Advance Design System (ADS) — Momentum 2.5D EM solver [14]. To validate the simulated results, the practical prototypes of the antennas were constructed. Fig.2 and 3 illustrates the computed and experimental results of the two antennas. Two adjacent resonant frequencies in the range of return loss > 10 dB are observed, i.e., 1.61 and 2.485 MHz. 'Ant1' shows the measured impedance bandwidth of 6.25% at 1.6 GHz and 6% at 2.475 GHz whereas the 'Ant2' exhibits narrower bandwidth of 3.15% at 1.6 GHz and 6% at 2.525 GHz. These results confirm that the antennas completely satisfy the desired L frequency band (1.61-1.624 GHz) and ISM/S band (2.4835-2.5GHz) band respectively.

Fig.4 and 5 depicts the axial ratio of the proposed antennas for y-z plane (φ = 90°). As can be noticed, an axial ratio of less than 3dB over ±45°elevation angle for 'Ant1' whereas it is less than 4 dB over ±40°elevation angle for 'Ant2'. The measured gains for two proposed antenna are shown in Fig.6. The measured gains for both antennas varied between 1.25 and 2.25 dBi over the entire L band; and between 1.4 and 2.75 dBi over the ISM/S band. These results are quite promising and encouraging for practical deployment.

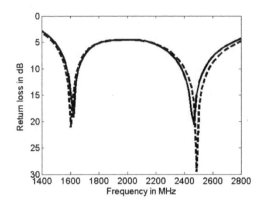

Fig. 2. Return loss of the proposed antenna (Ant1), where '————' simulated, ---------
'measured'

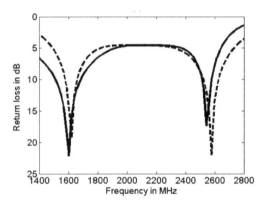

Fig. 3. Return loss of the proposed antenna (Ant2) , where '————' simulated, --------
'measured'

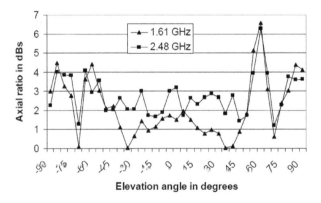

Fig. 4. Measured axial ratios for two operating frequencies versus elevation angle at $\varphi = 90°$ for
antenna (Ant1)

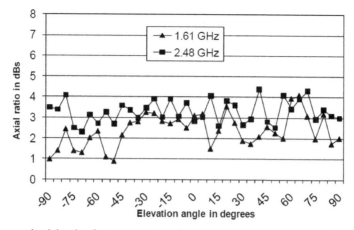

Fig. 5. Measured axial ratios for two operating frequencies versus elevation angle at $\varphi = 90°$ for antenna (Ant2)

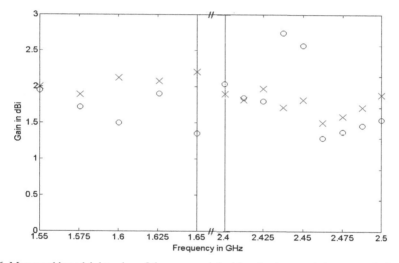

Fig. 6. Measured broadsight gains of the proposed dual-band antennas. (where xxxx is for Ant1 and ooo is for Ant2)

4   Conclusion

A new technique was discussed that reduced the antenna size for satellite-mobile handsets. Different stripline widths were introduced to achieve the frequency response and radiation performance required for Big LEO satellite mobile communications. Two different antenna geometries have been presented. The new designs were found to be very compact compared with four-square spirals on the same dielectric substrate described in previous studies. The larger antenna achieves a 6% impedance bandwidth in both bands, whilst the more compact design is limited to 3.15% in the lower band. The results in terms of the return loss are extremely promising and the field radiations are acceptable for both bands of interest, almost covering a ±45° range of elevation angles.

References

1. Dietrich, F.J., Metzen, P., Monte, P.: The globalstar cellular satellite system. IEEE Trans. Antennas and Propagation 46(6), 935–942 (1998)
2. Evans, J.V.: Satellite systems for personal communications. IEEE Antennas and Propagation Magazine 39(3), 7–20 (1997)
3. Daddish, S.M., Abd-Alhameed, R.A., Excell, P.S.: New Designs for Dual Band Antennas for Satellite-Mobile Communications Handsets. Applied Computational Electromagnetics Society Journal 15(3), 248–258 (2000)
4. Agius, A.A., Leach, S.M., Suvannapattana, P., Saunders, A.R.: Effects of the human head on the radiation pattern performance of the quadrifilar helix antenna. In: IEEE Int. Symp. On Antennas and Propagation, vol. 2, pp. 1114–1117 (1999)
5. Suvannapattana, P., Agius, A.A., Saunders, S.R.: Methods for minimizing quadrifilar helix antenna interactions with the human head. In: IEE Seminar on Electromagnetic Assessment and Antenna Design Relating to Health Implications of Mobile Phones, IEE Pub. No. 1999/043, pp. 11/1-11/4 (1999)
6. Ermutlu, M.E.: Modified quadrifilar helix antennas for mobile satellite communication. In: IEEE Conf. on Antennas and Propagation for Wireless Communications, pp. 141–144 (1999)
7. Wheeler, H.A.: Fundamental Limitations of Small Antennas. Proc. IRE 35, 1479–1484 (1947)
8. Khalil, K., Abd-Alhameed, R.A., Excell, P.S.: Dual–band quadrifilar square spiral antenna for satellite-mobile handsets. In: IEE International Conference on Antennas and Propagation, Exeter, vol. 1, pp. 186–189 (2003)
9. Zhou, D., Abd-Alhameed, R.A., See, C.H., Excell, P.S., Hu, Y.F., et al.: Quadrifilar Helical Antenna Design for Satellite-Mobile Handsets Using Genetic Algorithms. Microwave and Optical Technology Letters 51(11), 2668–2671 (2009)
10. Abd-Alhameed, R.A., Zhou, D., See, C.H., Excell, P.S.: Design of Dual-Band Quadrifilar Spiral Antennas For Satellite-Mobile Handsets. Microwave and Optical Technology Letters 52, 987–990 (2010)
11. Alazhari, K., Abd-Alhameed, R.A., Excell, P.S., Khalil, K.: New designs for single and dual band quadrifilar spiral antennas (QSA) for satellite-mobile handsets. In: IEE International Conference in Antennas and Propagation, Manchester, Manchester, vol. 2, pp. 750–753 (2001)
12. Nakano, H., Okuzawa, S., Ohishi, K., Mimaki, H., Yamauchi, J.: A Curl Antenna. IEEE Transactions on Antennas and Propagation 41(11), 1570–1575 (1993)
13. Colburn, J.S., Rahmat-Samii, Y.: Quadrifilar-Curl Antenna for the Big-LEO Mobile Satellite Service System. IEEE Antennas and Propagation Society International Symposium Digest 2, 1088–1091 (1996)
14. Momentum- 2.5D EM Simulator, Agilent Advance Design Systems, http://www.home.agilent.com/agilent/

Network-Based Mobility with DVB-RCS2
Using the Evolved Packet Core

Fabian A. Walraven, Pieter H.A. Venemans, Ronald in 't Velt, and Frank Fransen

TNO, Eemsgolaan 3, 9727 DW, Groningen, Netherlands
{fabian.walraven,pieter.venemans,ronald.intvelt,
frank.fransen}@tno.nl

Abstract. The network of the future consists of a combination of different access networks, each providing a level of network availability and mobility suited for a wide range of applications. Mobile network developments culminated in work on the E-UTRAN and Evolved Packet Core (EPC) network and can provide mobile broadband access to many citizens. However; the reach of these or other networks leaves remote and rural areas without network coverage, even in Europe. Satellite communication networks traditionally have a strong position in serving remote and rural areas and can potentially fill the gap. On the other hand, satellite communications can benefit from technology reuse by aligning with mobile standardization initiatives. The work presented in this paper describes how DVB-RCS2 satellite networks can be connected to the network-agnostic EPC to achieve network-based mobility with terrestrial mobile networks. We identify challenges and propose optimizations to improve the integration.

Keywords: LTE, DVB-RCS, EPC, mobility.

1   Introduction

The network of the future consists of a combination of different access networks, connected to an IP based core network, and each providing a level of network availability and mobility suited for a wide range of applications. The initiative of the 3rd Generation Partnership Project (3GPP) to define a competitive mobile network for the long term resulted in LTE or "Long Term Evolution". This standardisation effort includes both a new radio access network (evolved UMTS terrestrial radio access network, or E-UTRAN) [1] and a packet core network ("Evolved Packet Core", or EPC) [2] which supports mobile services over multiple access technologies, and has found broad industry acceptance.

EPC is a multi-access, multi-service architecture that includes policy and QoS control, charging, and service continuity functions under control of a core network operator. Supported access networks include both 3GPP access networks such as E-UTRAN and non-3GPP networks like WiMAX, WiFi and wired networks.

The EPC builds on established Internet standards from the IETF, applied in an operator environment. It supports both client-based and network-based mobility

P. Pillai, R. Shorey, and E. Ferro (Eds.): PSATS 2012, LNICST 52, pp. 87–94, 2013.

between 3GPP and non-3GPP access networks. Charging, policy and QoS control are supported through the Policy and Charging Control (PCC) architecture [3], which provides an access-independent framework to control and monitor access network resources.

In satellite communications, DVB-RCS [4] is the only multi-vendor VSAT standard for interactive service access. DVB-RCS2 [5] is the successor of DVB-RCS of which the lower and higher layer specifications were recently approved by the DVB Project, and formal standardization is planned through ETSI early 2012.

DVB-RCS2 provides improvements on all layers compared to the first DVB-RCS set of standards. It has native support for IPv6, improved performance, enhanced security and QoS control. The forward channel is based on the DVB-S2 specification [6] for efficient transport of IP traffic, and higher layer protocols draw heavily from IETF standards. These improvements and a standards-based approach, combined with wide coverage make DVB-RCS2 a viable access option in the future IP networks. For example, it can connect underserved remote and rural areas as a complement of terrestrial mobile networks. Sharing a common packet core network offering mobility management, secure communication and QoS control would aid integration and promote technology reuse.

The work presented in this paper describes how DVB-RCS2 satellite networks can be integrated with the EPC to achieve network-based mobility with terrestrial mobile networks. We identify challenges and propose optimizations to improve the integration, for which an illustrative scenario is considered (adapted from [7]): a dual-radio terminal serves as a network gateway towards a mobile backhaul (e.g., a backhaul for a WLAN in a train). When the terminal has cellular network coverage, it is connected to E-UTRAN; otherwise it performs a handover to a DVB-RCS2 network.

First we describe network-based mobility features of the EPC in section 2, followed by the policy and control features in DVB-RCS2 in section 3. Section 4 then describes the integration of the two networks, and we conclude in section 5.

2   EPC and Network-Based Mobility

A high-level overview of the EPC is depicted in **Fig. 1**, including both 3GPP and non-3GPP access networks. Non-3GPP accesses can be "trusted" or "untrusted", where untrusted access networks require the operator to deploy an evolved packet data gateway (ePDG) that ensures authentication of the user equipment (UE) and secure access to the EPC. For trusted accesses the EPC relies on the access network for providing appropriate security mechanisms. A packet data network gateway (PDN GW) is responsible for IP address assignment, and connects to a serving gateway (SGW) and access gateway (AGW) in the 3GPP and non-3GPP access networks, respectively. The policy and charging rule function (PCRF) is part of the PCC architecture and is a policy decision point for QoS enforcement and gating by both the policy and charging enforcement function (PCEF) in the PDN GW and the bearer binding and event reporting function (BBERF) in the access network gateways. For profile access and authentication, authorization and accounting (AAA) the home subscriber (HSS) and 3GPP AAA servers are used.

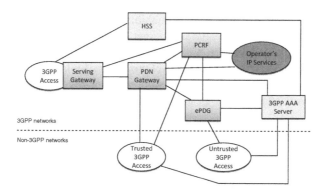

Fig. 1. The EPC architecture [8]

The EPC includes two mobility protocols for handovers between 3GPP and non-3GPP access networks: DSMIPv6 [9] for client-based mobility and PMIPv6 [10] network-based mobility. In this paper we focus on network-based mobility management.

The PDN GW implements a local mobility anchor (LMA). In the access network there is a functionality called the mobile access gateway (MAG). The MAG sends binding updates to the LMA on behalf of the UE. When the UE attaches to another access network the MAG makes sure the UE keeps the same IP address and that the LMA knows where the UE is now attached. To be able to uniquely identify a UE across multiple access networks (each could implement its own identification mechanism), 3GPP has introduced a network-access identifier (NAI) based on the international mobile subscriber identity (IMSI) for all PMIPv6 interfaces.

The 3GPP standards describe how a handover is performed from a 3GPP to a trusted non-3GPP access network using PMIPv6. First the UE is attached to a 3GPP access network (e.g., E-UTRAN). When a supported trusted non-3GPP access network is discovered the UE can initiate a handover based on mobility policy rules. The handover starts with non-3GPP attachment procedures, after which the authentication procedure starts based on the improved extensible authentication protocol method for 3rd generation authentication and key agreement (EAP-AKA') [11]. When the UE is authenticated, the UE can trigger an attachment on the network layer (IP), for instance a DHCP request. This triggers the policy control system in the trusted non-3GPP access network to contact the PCRF in the EPC. After this step the PMIPv6 proxy binding update is sent by the MAG in the trusted non-3GPP access towards the LMA in the PDN GW, and the policy control function in the PDN GW contacts the PCRF. The address of the associated PDN GW is registered in the HSS, and the LMA sends a proxy binding acknowledgement towards the MAG. At this moment the PMIPv6 tunnel is created. Optionally, extra policy rules can be provisioned in the trusted non-3GPP access network. This concludes the network layer attachment of the UE which now has an IP address assigned, and the connection to the 3GPP access network is released.

Note that for the support of network mobility (i.e. mobility of not just one host but an entire network segment for which the UE is the network gateway) work is ongoing in the IETF on NEMO [12], which is an extension of the DSMIPv6 (client mobility) standard that not only delivers address mobility for one host, but does this for a complete IP network. More recently an Internet Draft was published [13] describing the use of NEMO-like capabilities in combination with PMIPv6.

Now that we have described the high-level procedures for handovers using network-based mobility in the EPC, the following section elaborates the DVB-RCS2 network focusing on network attachment to be able to perform a handover using the EPC.

3   DVB-RCS2

DVB-RCS2 can be used in different network topologies of which the transparent star topology is most common. A satellite network operator can divide its network into one or more virtual networks which are assigned to one or more virtual network operators. An active RCS terminal (RCST) can be a member of only one virtual network. For simplicity we consider a DVB-RCS2 network relative to such virtual network (i.e. we do not consider the existence of multiple (virtual) operators in the same physical network). The RCST connects to a network control center (NCC) to get access to and request resources from the satellite network. The NCC manages the network resources and distributes the real-time network configuration using the DVB-S2 forward link. The transparent gateway (TS-GW) is the function forwarding user traffic to and from the satellite network (often collocated with the NCC), and the network management center (NMC) provides overall management of the network elements (including service level agreements assigned to a RCST). The DVB-RCS2 specification is split in a lower layers (MAC and below) and higher layers part. Below we discuss network attachment and QoS procedures in a DVB-RCS2 network relevant for the handover procedure.

When a RCST wants to attach and logon to the network it first initializes the physical layer such that it can receive network information and configuration from the NCC through DVB-S2. When the information is received and processed by the RCST it can send a logon request to the NCC. It then enters a negotiation phase which on success will complete initialization of the lower layers and puts the RCST in an operational state (called the TDMA Sync state). Typically 2 to 4 protocol exchanges are needed to get to this state which completes MAC layer attachment, each exchange incurring latency of around 500ms due to propagation delays. When the logon is accepted the RCST receives several descriptors in the terminal information message (TIM) sent by the NCC: for example a descriptor with unsolicited timeslot allocations, or a list of request classes that describe lower layer services and their restrictions.

The RCST can request additional resources using solicited allocation which are mapped to request classes. It can use these network resources for higher layer initialization (e.g., request for an IP address using DHCP).

The RCST uses an IP classification table and higher layer service (HLS) mapping table to map IP traffic to request classes. These tables are provisioned by the NMC using management commands (not part of the logon procedure). A higher layers initialization descriptor is used to be able to boot the higher layers by the NCC at logon. For user traffic interfaces a DHCP option descriptor can be included in a TIM which for each MAC interface can provide configuration information. The supported DHCP options can be advertised by the NCC in a TIM broadcast (TIM-B) message received by all terminals (as opposed to a TIM unicast or TIM-U message which is designated to a specific terminal). The RCST requests for specific DHCP options in the logon request. A subsequent higher layer DHCP exchange is needed to obtain an IP address for the interface.

In an operational state (after a successful logon) the NCC can force the RCST to disable transmission, after which the RCST enters a standby state. The forward link is kept in an operational state during standby, while the RCST ceases transmission and associated network resources are released. In order for the RCST to return to an operational state a new logon procedure is required (e.g., after an enable transmission instruction from the NCC). These procedures can be used by the NCC to control the RCST before and during a handover.

4   Handover to a DVB-RCS2 Access Network

In geostationary satellite communication networks resources are relatively expensive and latencies high compared to terrestrial mobile networks (mostly due to propagation delays). We consider two prerequisites for integration: the RCST should only attach to the network when DVB-RCS2 becomes the active access network (according to the mobility policy), and the number of control protocol messages exchanged over the DVB-RCS2 network to perform the handover should be minimized. A trade-off following from these prerequisites is the state in which the RCST is kept when it is inactive: keeping a terminal attached makes IP address preservation across handovers difficult, and can result in unused network resources (especially when a large share of resources is unsolicited), however it would allow for a faster handover procedure (because less protocol exchanges result in less handover delay). In our proposed solution (shown in **Fig. 2**) we try to strike a balance between these two modes, considering both mobility and QoS management.

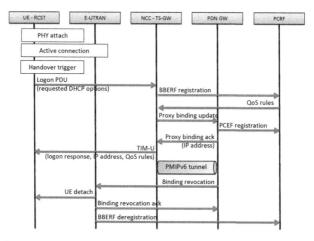

Fig. 2. Optimized handover to the DVB-RCS2 access network using the EPC

4.1   Attachment and Mobility Management

Network selection in the EPC can either be manual or through mobility policies. In our illustrative scenario a handover takes place automatically based on changing network conditions (e.g., when a terminal moves out of E-UTRAN coverage). We assume that policies are pre-provisioned for the UE to avoid extra control protocol exchanges. The pre-provisioned policy contains rules that trigger a handover to the DVB-RCS2 network based on E-UTRAN network availability. The triggers for a handover can be access-network specific and are out of scope of the EPC standards; in any case when a drop in network signal occurs the satellite network should establish connectivity before network connectivity to the E-UTRAN is lost and disruptions become noticeable. This could for example be achieved with signal strength measurements and thresholds, or a provisioned coverage map combined with real-time location information.

In a dual radio setup the delay introduced by attachment to the DVB-RCS2 network could be compensated by maintaining connectivity to the 3GPP access network until attachment is complete (i.e. *"make before break"*); however this is of limited use when a fast moving terminal (e.g., on the train from our illustrative example) moves out of network coverage and the radio signal drops quickly; in any case we believe there is a need for a procedure considering both handover delay and network resource efficiency for satellite networks.

Before a network handover is initiated based on mobility policy rules, the RCST is in a stand-by state such that (unsolicited) network resources are not committed to the terminal and IP address preservation is possible.

To reduce the number of control protocol exchanges during attachment we propose the use of a new DHCP option which indicates that a handover takes place and allows the RCST to retrieve its IP address in a TIM-U response (the IP address is assigned to the RCST by the PDN GW based on PMIPv6 procedures). Consequently the address is configured before IP-level connectivity is available for stateless [14] or stateful [15] address autoconfiguration. The NCC can announce the support of such DHCP option in the TIM-B message, and if supported by the RCST it can request for it in the logon request to the NCC. This would allow for attachment to layer 3 without additional higher layer control protocol exchanges (each exchange would cost at least 500ms), but requires the NAI to be present in the logon request.

Another consideration affecting handover attachment is network security. When the DVB-RCS2 network is assumed to be a trusted network to the EPC (note that this is not a characteristic of the access network but an operator agreement), the EPC relies on the access network security in the non-3GPP access network (i.e. DVB-RCS2 data link layer security). For the security between the non-3GPP access network and the EPC network domain security from 3GPP should be applied, as specified in [16]. This authentication would require additional control protocol exchanges before the handover is completed. Optimizations could be made by exchanging (part of) the authentication vector over E-UTRAN, or combine EAP exchanges with the DBV-RCS2 logon procedure.

4.2   QoS Management

Part of the handover is informing the EPC QoS functions of the access network change, which is separate from the PMIPv6 signaling. Although support of PCC is optional for trusted non-3GPP accesses, we do believe support is meaningful as it allows consistent QoS management across access networks. During a handover PCC integration involves registering the BBERF (located in the DVB-RCS2 TS-GW) with the PCRF, and the PCEF (located in the PDN GW) modifying the access network session by updating the PCRF. In response the BBERF receives the QoS rules and event triggers that it should enforce. The QoS rules contain maximum and guaranteed bitrates for defined service data flows (i.e. an aggregate set of packet flows).

In DVB-RCS2 QoS rules are not only provisioned in the TS-GW but also in the RCST, and we suggest to translate the PCC QoS rules into the request classes configuration in the lower layer service descriptor in the TIM-U in order to provide QoS rule enforcement in both the RCST and TS-GW: the RCST uses its IP classification table (now associated to the PCC service data flow templates in the QoS rules) and HLS mapping table to map IP traffic to the request classes. When the service data flow templates provided by the PCRF are changed this requires updating of the IP classification table and HLS mapping table using management commands by the NMC, therefore keeping these templates stable would improve performance of the handover with integrated QoS management.

5   Conclusion and Future Work

The integration of DVB-RCS2 with the EPC can strengthen the network of the future where multiple access networks provide connectivity, especially for remote and rural areas. Effort is needed on mobility management and QoS management to ensure an efficient integration given unique properties of geostationary satellite communication networks (e.g., large coverage, propagation delays). We have translated this observation in two prerequisites in our study of network-based mobility: the RCST should only attach to the network when DVB-RCS2 becomes the active access network, and the number of control protocol messages exchanged over the DVB-RCS2 network to perform the handover should be minimized. Based on these prerequisites we have proposed solutions to reduce the number of messages needed during a handover by adding a new DHCP option carrying an assigned PMIPv6 address in response to a successful DVB-RCS2 logon procedure. The same TIM-U response message is used for carrying QoS configuration to the terminal such that integration with PCC is established without requiring additional control protocol exchanges provided that some pre-provisioning is performed. Without optimization the handover would require around 8 to 10 protocol exchanges for authentication and attachment (both MAC and IP), which can be reduced to 4 or less. Considering propagation delay of 250ms this shows that the integration of DVB-RCS2 with the EPC is feasible; however there is room and need for improvements to optimize network-based handovers.

Based on the proposed solution we are working on an implementation and validation of the proposed solution in our EPC testbed and plan to propose enhancements that further improve the integration of satellite communication networks with terrestrial mobile networks.

References

[1]   3GPP TS 36.300: E-UTRAN Overall Description
[2]   3GPP TS 23.401: GPRS Enhancements for E-UTRAN access
[3]   3GPP TS 23.203: Policy and Charging Control Architecture
[4]   ETSI EN 301 790: Interaction Channel for Satellite Distribution Systems
[5]   DVB A155-1: Second Generation DVB Interactive Satellite System Part 1: Overview and System Level Specification
[6]   ETSI EN 302 307: Second Generation Framing Structure, Channel Coding and Modulation Systems for Broadcasting, Interactive Services, News Gathering and Other Broadband Satellite Applications (DVB-S2)
[7]   Cano, M., Norp, T., Popova, M.: Satcom Access in the Evolved Packet Core. In: Venkatasubramanian, N., Getov, V., Steglich, S. (eds.) Mobilware 2011. LNCS, Social Informatics and Telecommunications Engineering, vol. 93, pp. 107–118. Springer, Heidelberg (2012)
[8]   3GPP TS 23.402: Architecture Enhancements for Non-3GPP Accesses
[9]   IETF RFC 5555: Mobile IPv6 Support for Dual Stack Hosts and Routers
[10] IETF RFC 5213: Proxy Mobile IPv6
[11] IETF RFC 5448: Improved Extensible Authentication Protocol Method for 3rd Generation Authentication and Key Agreement (EAP-AKA')
[12] IETF RFC 3963: Network Mobility (NEMO) Basic Support Protocol
[13] IETF draft-ietf-netext-pd-pmip-01: Prefix Delegation for Proxy Mobile IPv6
[14] IETF RFC 4862: IPv6 Stateless Address Autoconfiguration
[15] IETF RFC 3315: Dynamic Host Configuration Protocol for IPv6 (DHCPv6)
[16] 3GPP TS 33.210: Network Domain Security (NDS); IP Network Layer Security

Handover Management for Hybrid Satellite/Terrestrial Networks

Fabrice Arnal, Riadh Dhaou, Julien Fasson, Julien Bernard, Didier Barvaux,
Emmanuel Dubois, and Patrick Gélard

ENSEEIHT / IRIT (Institut de Recherche en Informatique de Toulouse)
{riadh.dhaou,julien.fasson}@enseeiht.fr
Thales Alenia Space
fabrice.arnal@thalesaleniaspace.com
Viveris Technologies
{julien.bernard,didier.barvaux}@toulouse.viveris.com
Centre National d'Etudes Spatiales
{emmanuel.dubois,patrick.gelard}@cnes.fr

Abstract. Initially envisaged to support handover between different wireless
802.x network technologies, the IEEE 802.21 standard also appears as the good
candidate for handover management in future integrated satellite / terrestrial
systems. This paper presents an analysis of how this standard could be
implemented in the frame of a realistic scenario and taking into account the
current trends in wireless networks and mobility architectures. Our solution is
then evaluated by means of emulation over a DVB-RCS representative testbed,
and based on an experimental MIH implementation. We finally show that
seamless handover can nearly be achieved with very short service outages.

Keywords: MIH, Satellite, 3GPP, LTE, Handover, Mobility, Test-bed.

1 Introduction

These last years show a tremendous evolution of GEO systems towards the
integration of terminal mobility. This association with wireless terrestrial networks is
a relatively new idea. Several proprietary hybrid systems integrating GEO satellites
and ground components arise. The design of these integrated/hybrid systems takes
into account physical, MAC, and network layers issues [1]. The combination of these
two technologies can achieve for broadband services full coverage and high capacity.
However the convergence at network level is still an open issue.

The main networking trial is certainly seamless handover. Managing handover in a
couple of heterogeneous technologies may rely either on specific optimised or on
standard and generic solution such as Media Independent Handover (MIH). The first
solution may be optimised for each case of technologies. The last solution evolves
more naturally and easily in order to integrate new access technologies. However
there is a clear lack of experience in the deployment of this kind of solution.
Therefore, our position here is to propose realistic hybrid network architecture and to
see how MIH could be integrated.

P. Pillai, R. Shorey, and E. Ferro (Eds.): PSATS 2012, LNICST 52, pp. 95–103, 2013.
© Institute for Computer Sciences, Social Informatics and Telecommunications Engineering 2013

The paper is organized as follow: Section 2 presents existing works related to mobility and MIH backgrounds. Section 3 identifies a possible use case for satellite/terrestrial integration and describes a relevant network architecture and design choices for the MIH usage. At last, Section 4 presents the evaluation of this solution through experimentations lead on an emulation platform.

2   Inter-system Mobility Architectures and MIH Backgrounds

The major standard organisations support mobility features:

- **IETF** proposes network, transport and session level mobility solutions, therefore usable for inter-system handover. The MIPSHOP and MEXT working groups focus on mobility issues in heterogeneous networks. Several mobile IP based solutions and extensions were proposed and a big trend has been raised on the development of "Smart Routers" products that are based on such solutions. Transport level mobility solutions are based either on TCP (MP-TCP, Freeze-TCP and Mobile Socket Service) or on SCTP. SIP-based mobility was also proposed.
- **3GPP/3GPP2** architecture not only supports inter-technology mobility but also vertical handover (e.g. with WiFi and Wimax). Two operation modes are defined: the non-optimized and the optimized handovers. For instance, VoIP services, in a 2G/3G network, are based on the first mode whereas no standardized solutions exist for the second one. The network level mobility is based on MIPv4, DSMIPv6 and PMIP (Proxy MIP).
- From another hand, **IEEE** has formed the 802.21 group, where MIH was formerly proposed for handover between 802.x access technologies. Later, the standard has been extended for 3GPP networks support.

MIH provides abstract services to the higher layers using a unified interface (L2.5 functionalities). MIH is not designed to take handover decision but rather provides the communication tools to assist the handover process. MIH defines three different services, i.e. Media Independent Event Service (MIES), Media Independent Command Service (MICS) and Media Independent Information Service (MIIS). MIES provides events triggered by changes in the link characteristic and status. MICS provides the upper layers necessary commands to manage and control the link behaviour to accomplish handover functions. MIIS provides information about the neighbouring networks and their capabilities. Extensive details on the IEEE 802.21 standard can be found in [2].

One major question is certainly the reason to use MIH in satellite context. Indeed, the needs of vertical handover schemes [3] have been shown in several architectures based on the 802.21 standard [4, 5, 6].

Existing Studies on MIH, Implementations and Testbeds
To our knowledge, there is no reference implementation of the IEEE 802.21 standard. Nevertheless, FP6/FP7 IST projects have contributed to a prototype on GNU/Linux [7]. Another prototype proposes an implementation of a mobile VoIP using SCTP

with automatic address reconfiguration [8]. An experimental and analytical study was conducted [9] to determine the adequacy of protocols for mobility networks. This study presents the handover procedures (based on MIPv6 and PMIPv6) in WIFI/CDMA networks. PMIPv6 based on reactive handover method has been also observed in [10]. In [11], a 802.21 client is simulated in a heterogeneous environment and the paper analyzed the effect of speed on terminal handovers between WiFi and 3G, but without validation of these simulations. Last, MIH has been proposed for future aeronautical communications [12].

The use of MIH was also discussed in several prototypes. An implementation of 802.21 with a limited number of features was presented in [13]. In [14] the integration of broadcast technology in heterogeneous networks using MIH is discussed. [15] presents a solution for handover between heterogeneous networks that integrates MIH and Mobile Proxy Agent by defining an API performing the MIH Service Access Point. Eventually there is an extended research activity on seamless handover. Models of MIH are integrated to simulation tools such as Qualnet [16] and ns-2 [17]. Open source implementations such as openMIH [18] and Odtone [19] have been also set up.

3   Scenario Description

Mobile Services

A hybrid system aims at offering access to users' services wherever they are by enlarging the system coverage. However, such systems only make sense if they offer a mobile service. For a single user to a whole group of users in the same vehicle, applications are miscellaneous, from voice applications to Internet access.

Access Networks

Between the different types of satellites, GEO systems offer the best opportunity to cope with the terrestrial part in terms of coverage and cost. Apart private solutions, the main standards proposing mobility for GEO are GMR, the mobile version of DVB Return Channel for satellites (DVB-RCS+M) and DVB for Satellite Handheld (DVB-SH) that focus on mobile broadcast services. GMR technology has been designed for extension of the cellular world to satellite and thus it focuses on voice services. Some integrated systems, like Terrestar [20] are nowadays using GMR technology to interconnect it to a mobile operator core network. Indeed GMR standard has evolved, following the "all-IP" wave. Therefore a LTE solution adapted to satellite appears as the right direction for GMR standard. In Ku/Ka bands, the use of DVB-RCS+M is highly relevant for vehicular services (i.e. mobile collective terminal) and this will be retained for our reference scenario, including for the test-bed implementation.

Concerning the terrestrial networks, we could consider any kind of terrestrial networks that may provide mobility. If LTE and Wimax are two promising solutions, the limits of test-bed have induced the use of a simpler technology, WIFI.

Network Architecture

A major element in network architecture is the different roles undertaken. We assume a mobile service provider (MSP) ensures the IP connectivity of its mobile clients (final users and/or hybrid terminal) and may provide some other kinds of services.

This role is generally linked to the core network of final user. MSP is in charge of interaction with the manager of the different access networks that its clients are using. If needed, it may rent bandwidth at user's demand. The following Fig. 1 illustrates a distribution of roles through a DVB-RCS+M / LTE architecture. A group of users are connecting through a LAN in a train to their services. The hybrid terminal can connect to an E-nodeB or to the satellite Gateway. Finally, a Mobile IP / NEMO architecture is chosen for network mobility support.

The terrestrial and satellite networks are access networks between the user and his MSP. The MSP may propose the first visible router to user and it manages the AAA functionalities. However, some of these functionalities can be delegated to access network, using MSP Point of Presence (PoP). Moreover, MSP role can directly be undergone by one of the two access networks, especially for the 3GPP one. This option is more convenient for LTE proposing a whole set of services like IP Multimedia Subsystem (IMS) as shown on the figure. For satellite systems or Wimax the core network is less developed than cellular one. So this integration demands more change to their network architectures.

Policing functionalities can be distributing on both sides and centralized by a global AAA entity in the MSP. In the case of a LTE network this function is managed by the Policy and Charging Resource Function (PCRF) whereas for others access networks the control at IP level can be made in implementing the IETF COPS standard. Policy Enforcement Point (PEP) and Policy Decision Point (PDP) are then introduced in the architecture, the former being deployed in the access network while the later is directly managed by the Mobile Service Provider (MSP). The interaction between these functionalities and the handover signaling is however out of the scope of this paper.

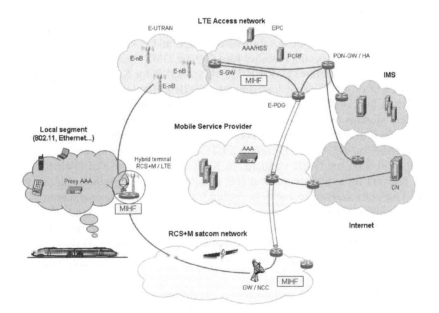

Fig. 1. An example of network architecture for the DVB-RCS+M/LTE reference scenario

Handover Decision
MIH supports both mobile-initiated handover as well as network-initiated handover, both alternatives can be preferred according to the exact deployment. In our scenario, the terminal chooses its mode of connection and the timing to perform the handover. A simple decision algorithm is implemented as an independent process (as a MIH user) in the terminal and based on predefined Signal-To-Noise Ratio thresholds. Thanks to this modular architecture, the handover decision and/or policies is open and could be easily extended with much sophisticated algorithms.

MIH Functionalities
The MIH Function (MIHF) aims at referencing any event and at informing its users. In our proposal, MIHF is present in mobile and network elements, as it is the most general solution. For Wimax, MIHF is integrated on the access points whereas for LTE it may be the PDN-GW. However if the MSP takes part in the handover decision, it must have an MIHF so as to communicate with the handover decision function. MIHF is communicating with other MIHF in the different networks and they collect information, inform or receive order from the MIH users on the same entities. Another interesting MIH user is the MIIS server which can be used to select the available networks. However this option has not been chosen here.

MIH Primitive Selection
Many primitives are defined in the MIH standard although all are not necessarily required to be implemented. The selection of the right primitives may allow a seamless handover. The selection has been led by the need to obtain a seamless handover and the convenicnce of the deployment for the tests. The following primitives have been used for handover indication:

- MIH_Link_Configure_Thresholds is used to fix the thresholds for signaling the need of a handover before the link is going done.
- MIH_Link_Parameters_Report proposes to inform the different entities when the threshold is reached. It allows the preparation of the handover.
- MIH_Link_Going_Down is used by the L2 interface to announce the near link outage.

4   Implementation

Description of the Emulation Test-Bed, Implementation and Scenario
The core platform used is the Platine emulation tool [21], representative of a DVB-S2/RCS access network and running under a Linux environment. NEMO was integrated to Platine from the UMIP v0.4 software upgraded with the NEPL patch [22]. Standard Linux Route Advertisements (RADVD) are used for address auto-configuration. Finally the ODTONE v0.2 beta stack was used for MIH implementation, modified by minor adaptations for our needs. In order to model the LTE/3GPP connection we have used a WIFI link which was more convenient for performing tests in our lab environment. In addition, only the handover from WIFI to

satellite could have been tested, as few extra developments were further needed for Satellite to WIFI handover in our platform.

The (emulated) mobile network is connected through a hybrid Satellite Terminal that can be connected to the satellite gateway through the emulated satellite link, or to the WIFI access point. The NEMO instance and the MIH local function are therefore deployed on the Satellite Terminal, while the satellite and WIFI gateways host the network-side MIH Functions.

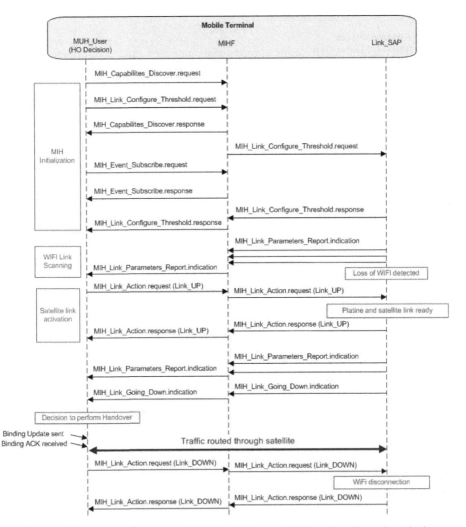

Fig. 2. MIH sequence diagram between the MIH user, MIH functionality and test-bed

Signal-to-Noise Ratio (SNR) variations are emulated by means of a pre-computed scenario stored in a file indicating the virtual SNR measurements. They are polled each 5 seconds by the mobile MIH Function (LINK_SAP interface). Two SNR thresholds are configured at the MIH stack (at the MIH startup, with LINK_CONFIGURE_THRESHOLD). Each threshold is mapped to a given MIH event (LINK_GOING_DOWN and LINK_PARAMETER_REPORT). The first event, when received, is used to start the Platine logging process. If successful, Route Advertisements can be received for the satellite connection through the virtual Platine network interface. Subsequently, when the second threshold is crossed, it triggers the NEMO handover phase. We capture this event to send an inter-process signal (SIGUSR2) to the NEMO stack in order it performs the Binding Update / Acknowledgement procedure. For future improvements, this interworking will be developed in a proper module that will also implement separately the Decision algorithm for flexible evolutions.

Application flows are composed of a CBR audio and a VBR video stream (RTP/UDP), each generated by VLC at a host located in the mobile network, and sent to a Correspondent Node (CN). This receiver is located at a neighboring network interconnected to both WIFI and satellite networks.

4.1   Results

A Wireshark client running at the CN measures the incoming rates. As shown in Fig. 3, a short outage (0.6 seconds) is experienced just after the handover at second 90. At the receiver, the audio flow starts to be corrupted, but the video quality remains intact.

Fig. 3. Incoming bit rates variations measured at Correspondent Node

Our investigation shows that no packet loss happens during the handover, which is a first point of satisfaction. With comparison to a basic scenario test conducted in the same conditions but without MIH, that cannot be presented here due the lack of space, a packet loss sequence lasting nearly for 4.5 seconds was observed.

Immediately after the handover, we understood in observing an increase of the satellite queue utilization that the introduced delay is not related to any mobility latency but to the satellite access layer allocations loop (Rate-Based Dynamic Capacity allocations were used here). In addition, the sudden propagation delay change (tenths of milliseconds with WIFI, 300ms with satellite) also adds a contribution to the observed delay. However since the video client largely buffers the receiving flow, this latency effect remains transparent, whereas audio samples are by far less buffered.

For the remaining period, audio remains corrupted and not video. This is due to the absence of any QoS support on the overloaded satellite connection. Buffering at the satellite link layer continues is never completely absorbed. Except this effect, we conclude this MIH integration is successful and achieves good performances, though only based on an experimental stack.

5   Conclusion

This paper has demonstrated that MIH, that appears complex at first glance, is actually well applicable to any hybrid IP satellite network, such as the future LightSquared system in the US based on LTE. Coupled with Mobile IP network mobility, it is able to show very good performances provided that the loss of signal can be detected before the link completely goes down. In our future works, derived scenarios with the support of QoS will be presented in an IMS environment, in order to illustrate how resources could be provisioned before the mobile switches to another network. We also intend to complete our integration work by supporting Satellite to WiFi handover, and maybe to implement handover decisions based on real SNR and/or location measurements.

References

1. Kota, S., Giambene, G., Kim, S.: Satellite component of NGN: Integrated and hybrid networks. Int. J. of Satellite Communications and Networking (May 2010)
2. IEEE 802.21 IEEE Standard for Local and metropolitan area networks— Part 21: Media Independent Handover Services (2009)
3. Márquez-Barja, J., et al.: An overview of vertical handover techniques: Algorithms, protocols and tools. Computer Communications 34, 985–997 (2011)
4. Salhani, M., Dhaou, R., Beylot, A.-L.: Terrestrial Wireless Networks and Satellite Systems Convergence. In: Int. Communications Satellite Systems Conference, AIAA (2007)
5. Qureshi, R., Dadej, A.: Adding Support for Satellite Interfaces to 802.21 Media Independent Handover. In: ICON (2007)
6. Hu, Y.F., et al.: Mobility Extension for Broadband Satellite Multimedia. In: IWSSC (2009)
7. Piri, E., Pentikousis, K.: Towards a GNU/Linux IEEE 802.21 Implementation. In: Proc. IEEE International Conference on Communications, ICC 2009, pp. 1–5 (2009)
8. Chen, Y.-M., et al.: SCTP-based handoff based on MIH triggers information in campus networks. In: 8th International Conference Advanced Communication Technology ICACT 2006, vol. 2, pp. 1297–1301 (2006)
9. Chiba, T., Yokota, H., Dutta, A., Chee, D., Schulzrinne, H.: Performance Analysis of Next Generation Mobility Protocols for IMS/MMD Networks. In: Proc. International Wireless Communications and Mobile Computing Conference, IWCMC 2008, pp. 68–73 (2008)
10. Choi, H.Y., Min, S.G., Kim, K.R., Han, Y.-H., Lee, H.B.: Seamless handover scheme for proxy mobile IPv6 using smart buffering. In: International Conference on Mobile Technology, Applications, and Systems, pp. 1–7. ACM (2008)
11. Melia, T., et al.: Analysis of effect of mobile terminal speed on WLAN/3G vertical handovers. IEEE Globecom (2006)

12. Ayaz, S., et al.: Architecture of an IP-based Aeronautical Network to Integrate Satellite and Terrestrial Data Links. In: ICNS (May 2009)
13. Dutta, A., et al.: Seamless Handover across Heterogeneous Networks - An IEEE802.21 Centric Approach. In: IEEE WPMC (2006)
14. Buburuzan, T., et al.: Integration of Broadcast Technologies with Heterogeneous Networks – An IEEE 802.21 Centric Approach. In: Consumer Electronics, ICCE (2007)
15. Tauil, M., et al.: Realization of IEEE 802.21 services and pre-authentication framework. In: Int. Conference on Testbeds and Research Infrastructures for the Development of Networks. Communities and Workshops TridentCom 2009, pp. 1–10 (2009)
16. Latkoski, P., et al.: SDL+QualNet: A Novel Simulation Environment for Wireless Heterogeneous Networks. In: SIMUTools 2010, Torremolinos, Malaga, Spain (March 2010)
17. Rouil, R., Golmie, N., Montavont, N.: MIH transport using cross-layer optimized stream control transmission protocol. Computer Communications 33, 1075–1085 (2010)
18. Lopez, Y., Robert, E.: OpenMIH, an Open-Source Media-Independent Handover Implementation and Its Application to Proactive pre-Authentication. In: Pentikousis, K., Blume, O., Agüero Calvo, R., Papavassiliou, S. (eds.) MONAMI 2009. LNICST, Social Informatics and Telecommunications Engineering, vol. 32, pp. 14–25. Springer, Heidelberg (2010)
19. Corujo, D., et al.: Using an open-source IEEE 802.21 implementation for network-based localized mobility management. IEEE Communications Magazine 49(9), http://helios.av.it.pt/projects/odtone
20. Vojcic, B., Mathesson, D., Clark, H.: Network of Mobile Networks; Hybrid Terrestrial-Satellite Radio. In: IWSSC 2009 (2009)
21. Baudoin, C., Arnal, F.: Overview of Platine emulation testbed and its utilization to support DVB-RCS/S2 evolutions. In: ASMS 2010, Cagliari, Italy, September 13-15 (2010)
22. http://software.nautilus6.org/NEPL-UMIP/index.php

Load-Aware Radio Access Selection in Future Generation Wireless Networks

Muhammad Ali, Prashant Pillai, and Yim Fun Hu

School of Engineering Design and Technology
University of Bradford, UK
{m.ali28,p.pillai,y.f.hu}@bradford.ac.uk

Abstract. In the telecommunication networks the introduction of Next Generation Wireless Networks (NGWN) has been described as the most significant change in wireless communication. The convergence of different access networks in NGWN allows generalized mobility, consistency and ubiquitous provision of services to mobile users. The general target of NGWN is to transport different types of information like voice, data, and other media like video in packets form like IP. The NGWNs offer significant savings in costs to the operators along with new and interesting services to the consumers. Major challenges in NGWN are efficient resource utilization, maintaining service quality, reliability and the security. This paper proposes a solution for seamless load aware Radio Access Technology (RAT) selection based on interworking of different RATs in NGWN. In this paper novel load balancing algorithms have been proposed which have been simulated on the target network architecture for TCP data services. The IEEE 802.21 Media Independent Handover (MIH) is utilized in load balancing specifically for mobility management, which enable low handover latency by reducing the target network detection time. The proposed method considers the network type, signal strength, data rate and network load as primary decision parameters for RAT selection process and consists of two different algorithms, one located in the mobile terminal and the other at the network side. The network architecture, the proposed load balancing framework and RAT selection algorithms were simulated using NS2. Different attributes like load distribution in the wireless networks and average throughput to evaluate the effects of load balancing in considered scenarios.

Keywords: NGWN, Load balancing, radio resource management, heterogeneous wireless networks, load balancing in wireless networks, vertical handovers, satellite-terrestrial wireless networks load balancing.

1   Introduction

Modern mobile devices like cell phones, PDA's, Tablet PCs already support multiple wireless technologies like UMTS, WLAN and Bluetooth and in the very near future would also support satellite and WiMax with multiple interfaces provision. While most of these devices are able to scan the different available networks the user would

P. Pillai, R. Shorey, and E. Ferro (Eds.): PSATS 2012, LNICST 52, pp. 104–112, 2013.
© Institute for Computer Sciences, Social Informatics and Telecommunications Engineering 2013

manually select which network he or she may want to use. It is envisaged that in the NGWN these devices may be able to apply some complex Radio Access Technology (RAT) selection techniques to find the most suitable network. Such a RAT selection technique may need to consider various parameters like received signal strengths, errors rates, costs, user preferences, QoS requirements, etc. Such a RAT selection technique would not only play an important part when a user switches on his or her mobile device but also when the user moves around. While most of the current day mobile networks already support seamless handovers, these are restricted to handovers within the same technology, i.e. horizontal handovers. It is envisaged that to efficiently use the network services the future mobile devices shall also support handovers across different radio access technologies. This process of switching mobile devices connectivity from one technology to another type of technology is called vertical handover. The joint call admission control (JCAC) algorithm for next generation heterogeneous wireless networks is envisioned as user-centric. User centricity implies that user's preferences are considered in decision making for RAT selection. However user-centric JCAC algorithms often lead to highly unbalanced networks load, which cause congestion on overloaded network and eventually increase the call blocking and call dropping probabilities. The unbalanced load situation in co-located networks also causes the poor radio resource utilization as some networks remain under loaded and some get over loaded. The load balancing strategies are required to efficiently utilize the available radio resources and avoid the unwanted congestion situations due to overloaded wireless networks.

This paper presents a novel NGWN RAT selection technique which uniformly distributes the network load between co-located heterogeneous wireless networks. It utilizes MIH to seamlessly handover mobile users between heterogeneous wireless networks for load balancing purpose. The advantage of this approach is that it minimizes the call blocking and dropping probabilities, number of packet drop/lost and delays during the handover process and enhances the network utilization by continuously balancing the load in co-located networks. The proposed load balancing approach monitors and controls the network load from both side (mobile node and network side), and addresses the most important problem in NGWN which is efficient resource utilization. The rest of the paper is organized as follows; section 3 describes the literature review of existing and presented load balancing RAT selection techniques, section 4 briefly describes the proposed load aware RAT selection algorithms and target network architecture. Simulation topology and results are discussed in section 5, which is followed by the conclusion.

2   Load Balancing Techniques

Usually more than one wireless networks may provide coverage to any given location in an urban area. For example, when working in an office building, the mobile device of a user may be in the coverage of a UMTS mobile network and a WLAN office network. A user may manually configure to use the UMTs network for voice services and the WLAN access for data services. In such overlapping coverage areas of different wireless networks such as satellite networks and terrestrial networks like WiMax, UMTS and WLAN; a RAT selection technique is required to find the most suitable network based on received signal strengths, errors rates, costs, user preferences, QoS requirements, and most importantly the load of networks.

The load balancing approaches presented in [2] and [3] have considered load balancing in homogenous network targeting WLAN. The approach in [2] considers the received signal strength indicator (RSSI) value to distribute the load between different access points (AP's) which have overlapping coverage areas. This approach uses the two values in balancing the load which are RSSI between mobile station (MS) and AP and the average RSSI value of all the MS's currently connected with AP. The method given in [3] considers both RSSI and the number of MS associated with AP which makes it much effective for load balancing. The technique used in [4] presented a solution for load balancing in homogeneous wireless networks, by utilizing genetic algorithm. As the genetic algorithm's convergence directly proportional to the size of population (mobile nodes and APs) therefore this approach is effective for WLAN networks and not for the heterogeneous wireless environment where population size is comparatively large due to large coverage areas. All approaches given in [2, 3, 4] were designed to enhance the performance for homogeneous network environment particularly WLAN.

In [5] load balancing approach has been presented which targets the proxy mobile ipv6 (PMIPV6) domain using MIH for heterogeneous networks. A comparison has been made between the scenario performing load balancing in extended PMIPV6 for handover signalling and the scenario using MIH signalling for load balancing. It was shown in the results that use of load balancing improves the efficiency whereas, MIH based load balancing improves data rate as compared to extended MIPV6 based load balancing. This disadvantage in this approach is when considering load-aware RAT selection; it is specifically designed for a MIPV6 architecture using Local Mobility Agent (LMA) and a new entity called Mobile Access Gateway (MAG) in the network. In [6] a general set of algorithms have been proposed which considers battery power of mobile users, received signal strength and load on available points of attachments in handover process to balance the load in co-located networks overlapping their coverage areas. In this approach load balancing is done only at network side without any interaction with the mobile node. On the other hand our proposed approach considers both; mobile nodes and network entities such as AP, BS and satellite ground station for load balancing thereby resulting in more efficient load balancing across the neighboring networks.

In [7] a detailed algorithm has been presented for network selection in heterogeneous wireless networks. The algorithm presented in [7] has been divided into two parts, one runs at mobile terminals and other part of algorithm runs at network entity such as basestation (BS) or access point (AP). This approach considers received signal-strength, battery power, speed, and location of mobile user but does not considers MIH which could have improved the handover process while moving the mobile nodes between different networks. In [8] a next generation networks (NGN) based approach has been presented in which hierarchical joint call admission control algorithm is extended to send newly added load reports from hierarchical call admission control (HCAC) entity to vertical call admission control entity (VCAC). The main goals of proposed approach in [8] are simplicity and scalability, however this approach performs balancing of load periodically and therefore may not performs very efficiently with abrupt load changes in different sub networks in the hierarchy. In [9] a Markov chain based model for load balancing and QoS based CAC has been presented and comparisons have been made between the results of load balancing based CAC and QoS based CAC algorithms. The load balancing approach presented in this paper more efficient than load based CAC approach presented in [9] as our approach uses MIH to minimize the handover delays when moving the mobile nodes for load balancing purpose and tends to uniformly distribute the load among available heterogeneous wireless networks.

3   Load-Aware RAT Selection Framework

3.1   Network Architecture

Figure 1 presents the target network architecture which is considered in this paper. It shows an MIH enabled multi interface mobile device which can use any of the three available wireless networks supported by its interfaces.

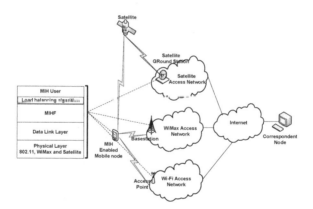

Fig. 1. Target network architecture

The access network of each technology such as Satellite, WiMax and Wi-Fi is connected to internet. There is also a correspondent node located behind the internet as shown in the Figure 1. The mobile node can communicate with the correspondent node over the internet using any available network which is supported by its interfaces. The mobile node handovers to different available networks while moving from coverage area of one network to another and during this mobility it can maintain the communication with correspondent node. The load balancing algorithms are located at the MIH user in MIH reference model as represented by the Figure 1. MIH user is selected for the load balancing process origin as MIH user is the central control point for triggering and handling MIH signalling as described in [1]. In mobile user the load balancing algorithm shown in Figure 2 is adopted and in Satellite Ground Station/BS/AP the load balancing algorithm for the network entity shown in Figure 3 is adopted.

3.2   Load Balancing Algorithms

This section describes the proposed load-aware RAT selection algorithm. The proposed algorithm considers the network type, signal strength, data rate and network load as primary decision parameters for RAT selection process and tries to maintain the load equilibrium on all networks which have common or overlapped coverage areas. It is assumed that all considered networks and mobile nodes support the IEEE 802.21 MIH.

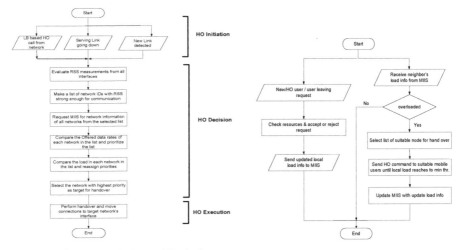

Fig. 2. LB algo. in the mobile device   **Fig. 3.** LB algo. at the Network side

The IEEE 802.21 MIH standard has been brought into use for seamless vertical handover operations of mobile nodes between the co-located wireless networks. The proposed approach has taken advantage of MIH media independent information service (MIIS) specifically for exchange of network load information besides exchanging other network related information like link type, link data rate, link capability, offered security and QoS and cost[1]. The proposed RAT selection framework consists of two load aware algorithms, one running on mobile device and other running on network entity like Satellite Ground Station, WiMax BS or WLAN AP. The flow chart shown in Figure 2 represents the proposed algorithm's which runs at mobile device. At the mobile device, the proposed technique first makes a list of available network IDs which are visible to mobile device such that received signal strength from those networks is higher than the minimum threshold. In next step load value of each network in the list is obtained from MIIS and compared. Then in following step it compares the data rate offered by each network in the list. The most preferred network from the list is the one with lowest load and highest offered data rate. The second algorithm shown in Figure 3 runs in network side. In the network entity like BS or AP the load balancing algorithm continuously keeps on updating the MIIS about its current load status and receives load information of its neighboring networks. This updating process runs on every time when a new connection starts or ends in the network. The most loaded network entity start moving out the suitable mobile users to appropriate networks, if the load variation is gone higher than threshold of 50% free resources margin, such that the percentage of free resources in one network is greater than or equal to the double of available resources percentage at any other network. Load balancing algorithm keeps on migrating out the suitable mobile nodes from over loaded network to the least loaded networks until the load in over loaded network becomes equal to or lesser than the average load in all the neighboring networks of overloaded network. The load balancing is performed by the handover procedures in both mobile and network side. In mobile nodes it is supported by the mobile node initiated handovers and on network entity it is supported by the network initiated handovers of the selected mobile nodes in the network.

4   Simulation Architecture and Results

4.1   Simulation Architecture

Figure 4 presents the simulation topology considered in this paper. Purpose for considering particular topology for simulation is to observe the effects of load balancing in most ideal scenarios where mobile nodes can see maximum overlapped coverage areas from different networks. Each mobile user maintains a TCP connection with the TCP source shown in Figure 4 throughout the simulation such that effects of handovers on active connections can be measured.

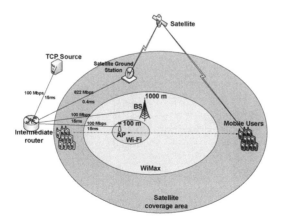

Fig. 4. Network topology for simulation

The scenarios considered in this paper consist of a group of mobile users which travel across the coverage areas of all three networks such as Satellite, Wi-Fi and WiMax as shown in Figure 4. In simulation a group of mobile users starts from the Satellite coverage area and move together towards the WiMax coverage area. At time 20 seconds all mobile users enter in WiMax coverage area and at approximately at 102 seconds they leave WiMax coverage area. The Wi-Fi coverage area is overlapped by WiMax therefore at time 62 seconds group of mobile users enters the Wi-Fi coverage area and at approximately 73 seconds all mobile users leave Wi-Fi coverage. Satellite coverage is available to the mobile users throughout the simulation time from time 0 seconds to 150 seconds. The TCP source shown in the Figure 4 maintains a TCP connection with each mobile node throughout the simulation.

4.2   Results

The simulation scenario discussed in the previous section is simulated using both load balancing and non-load balancing algorithms using the network simulator NS2 [10]. Results of average throughput and load distribution at different networks such as satellite, WiMax and Wi-Fi networks are shown in figures from Figure 5 to Figure 8.

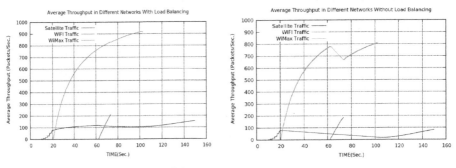

Fig. 5. Avg. Throughput with LB   **Fig. 6.** Avg. throughput without LB

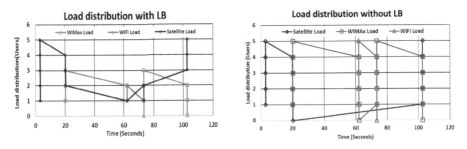

Fig. 7. Load distribution with LB   **Fig. 8.** Load distribution without LB

The Figure 5 is representing the average throughput graph of each network using load balancing algorithm, whereas the average throughput of each network using non-load balancing algorithm is shown in the Figure 6. It can be easily seen from the two graphs shown in Figure 5 and Figure 6 that with load balancing algorithm all the networks showed improved average throughput as compared to the non-load balancing algorithm for RAT selection. The load distributions shown in Figure 7 and Figure 8 represent the load status of each network throughout the simulation time. Figure 7 represents the load status when load balancing algorithm is applied and Figure 8 represents the load status of each network when load balancing algorithm is not applied. These results are showing that in scenario where load balancing algorithm is applied, the overall network load of each network remained lower as compared to the scenario where load balancing algorithm is not applied. As the load balancing algorithm divides the mobile users equally among the network having overlapped coverage areas and non-load balancing algorithm selects the network with highest data rate and highest signalling strength. Balancing the network load enable us to keep the network availability to maximum reducing the call blocking and dropping probabilities and utilizing the available resource efficiently.

5   Conclusion

In this paper a load aware RAT selection algorithm has been presented with the comparison of results generated by simulation scenarios using load balancing algorithm and without load balancing. Considered attributes for observation are load distribution on each of the network and average throughput of each network such as satellite, WiMax and Wi-Fi networks. The results showed that with load balancing both parameters showed improvement in the target hetcrogeneous wireless network architecture. The average throughput with load balancing is higher for each network as the overall load is divided by load balancing algorithm to avoid the congestion. Load balancing algorithm assures the fair load distribution between the overlapping networks whereas without load balancing different networks show abrupt load variations which decrease the performance with high congestion, high call dropping probability and blocking probability at overloaded network. Load balancing approach utilizes the available radio resources efficiently. Handover latencies are minimized, as it does not require all the mobile users to handover when load balancing algorithm is used. Hence the load aware RAT selection is a better approach as it offers high radio resource utilization with minimum number of handovers and hence low handover delays, minimized call/connection blocking and dropping probability and ability to maximize the network availability with uniformly distribution of load in co-located networks.

References

1. IEEE Std. 802.21, Media Independent Handover Services (2009)
2. Sheu, S.-T., Wu, C.-C.: Dynamic Load Balance Algorithm (DLBA) for IEEE 802.11 Wireless LAN. Tamkang Journal of Science and Engineering 2 (1999)

3. Papanikos, I., Logothetis, M.: A Study on Dynamic Load Balance for IEEE 802.11b Wireless LAN. In: 8th International Conference on Advances in Communications and Control, COMCON 2001 (June 2001)
4. Scully, T., Brown, K.: Wireless LAN load balancing with genetic algorithms. Knowledge Based Systems (2009)
5. Kim, M.-S., Lee, S.: Load balancing and its performance evaluation for layer 3 and IEEE 802.21 frameworks in PMIPv6-based wireless networks. Wireless Communications and Mobile Computing (2009), doi:10.1002/wcm.832
6. Lee, S., Sriram, K., Kim, K., Kim, Y.H., Golmie, N.: Vertical Handoff Decision Algorithms for Providing Optimized Performance in Heterogeneous Wireless Networks. IEEE Transactions on Vehicular Technology (January 2009)
7. Kaloxylos, A., Modeas, I., Georgiadis, F., Passas, N.: Network Selection Algorithm for Heterogeneous Wireless Networks: from Design to Implementation. Network Protocols and Algorithms (2009) ISSN: 1943-3581
8. Suleiman, K.H., Anthony Chan, H., Dlodlo, M.E.: Load Balancing in the Call Admission Control of Heterogeneous Wireless Networks. In: IWCMC 2006, Vancouver, British Columbia, Canada, July 3-6 (2006)
9. AL Sabbagh, A.: A Markov Chain Model for Load-Balancing Based and Service Based RAT Selection Algorithms in Heterogeneous Networks. World Academy of Science, Engineering and Technology 73 (2011)
10. http://www.isi.edu/nsnam/ns/index.html

Issues and Solutions When Deploying VPNs over Satellite Links

Dirk Gómez and Eriza Fazli

TriaGnoSys GmbH
Argelsrieder Feld 22, 82234 Wessling, Germany
{dirk.gomez,eriza.fazli}@triagnosys.com

Abstract. This paper summarises the issues created when deploying a virtual private network in broadband satellite systems. These issues are related to protocol enhancing, overhead, fragmentation, mobility, quality of service and network address translation. Solutions for these issues are proposed from the point of view of the satellite operator, since most depend on what degree of control it has over the network. More specifically, this paper explains which solutions can be applied when the satellite links an airplane with the ground.

Keywords: Security, VPN, IPsec, TLS, Satellite, Link, PEP, Fragmentation, Mobility, QoS, Aeronautical communications.

1   Introduction

Security in communications can be achieved with a private network that is only accessible by trusted members. However, nowadays traffic between two remote points goes through the public communications infrastructure to avoid the expensive investment needed to deploy a private infrastructure.

A Virtual Private Network (VPN) establishes a private communication over public infrastructure. For that, some protocols like IPsec and TLS/SSL are used. IPsec is a security standard specified by the IETF. It is designed to provide interoperable, high quality, cryptography-based security for IPv4 and IPv6. The set of security services offered includes integrity, authentication, protection against replays, and confidentiality. These services are provided at the IP layer, offering protection for IP and upper layer protocols.

The most common representatives of transport layer security are Secure Sockets Layer (SSL) and its successor Transport Layer Security (TLS). TLS/SSL is mainly used on top of TCP, which means that the TCP payload is protected but not the TCP header. The TLS protocol provides security in the form of reliability, integrity, anti-replay and (optionally) confidentiality.

The use of VPNs over satellite links creates additional issues to the ones already present in satellite links, like high delay, bandwidth-delay product and high error rates. These issues and solutions are summarised in the study described on this paper. For example, there are some known solutions like Multi-Layer IPsec to allow the use of transport protocol enhancing proxies inside the IPsec tunnel [1].

P. Pillai, R. Shorey, and E. Ferro (Eds.): PSATS 2012, LNICST 52, pp. 113–118, 2013.
© Institute for Computer Sciences, Social Informatics and Telecommunications Engineering 2013

From the satellite operator's point of view, the issues can be seen in three different cases, depending on which nodes it controls along the end-to-end communication path. The feasibility of the technical solutions shall later be then assessed with respect to the constraints posed by these control cases.

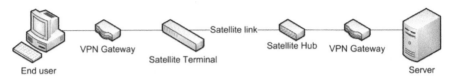

Fig. 1. The communication path as considered for the control cases

Three control cases are defined. Case 1 is when the satellite operator has control over the whole path. In Fig. 1, that corresponds to controlling from one VPN gateway to the other. Case 2 is when there is only control over the satellite related elements; satellite terminal, link and hub in Fig. 1. Case 3 is when, in addition to the satellite elements, one of the VPN gateways is controlled.

This paper starts by explaining the different scenarios considered and further details one of them as an example. Then, the issues of Protocol Enhancing Proxies, IP fragmentation, overhead, mobility, and quality of service are explained. More issues have been identified but are left out due to space limitation.

While the issues and solutions summarised are known, this paper's contribution is the analysis of the applicability of such solutions from the perspective of the satellite operator, according to the different situations represented by the previously defined control cases.

2   Scenarios

Different scenarios have been considered so that it would be possible to investigate: i) the three different control cases, ii) the two security protocols in scope, TLS and IPsec, and later iii) the feasibility of the technical solutions in real-life implementation scenarios. These include:

- **Public safety.** It describes de communications between emergency teams deployed on the field and their headquarters.
- **Aeronautical communications.** This scenario represents the communications between a mobile network and a static network. In this case, the mobile network is an aeroplane.

The aeronautical scenario is composed of two types of traffic, namely:

- **Safety-related communications** consist of communications between pilots and air traffic controllers (Air Traffic Services ATS) and communications between the aircraft and its airline (Airline Operation Control, AOC). The air-to-ground data communication is assumed to be provided by an entity called the Air

Communications Service Provider (ACSP). Because the satellite operator has some influence over the ASCP, this communication path is considered as having control case 1. The traffic in this path is mainly made of short messages, and it uses IPv6 taking into account the recent development in future air traffic management systems [2].

- *Non safety-related communications* consists of the traffic that comes from the passengers connecting to the internet (Airline Passenger Communications, APC) and forms the non-safety communications. Because there is no control at the ends of the communication (passenger and the internet) this is a case 2. IPv4 is considered in this path as it is still the most widely used protocol in the Internet.

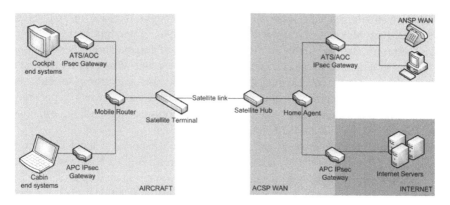

Fig. 2. Aeronautical communications scenario

3 Investigated Technical Issues

3.1 Protocol Enhancing Proxies (PEP)

Standard TCP might not work efficiently in satellite networks due to the bandwidth-delay product and the high bit error rate [3]. PEPs or enhanced TCP are usually used to improve the performance in such networks.

An enhanced version of a protocol differs from the standard protocol in the configuration of some parameters or behaviours (e.g. congestion control mechanism); it is a substitute for the original protocol. On the other hand, PEPs are new elements in the topology configuration that improve the performance of the protocol.

Some PEPs generate TCP packets as if they were the end hosts. Therefore, they must be capable of reading the fields in the TCP header. However, with IPsec these might be encrypted. Even if null encryption is used, after the PEP would generate a new packet, it would be unable to perform IPsec integrity protection and so, the new packet would be dropped at the IPsec gateways. Therefore, the PEPs should be placed outside the IPsec tunnel, requiring a control case 1 when an IPsec VPN is deployed. Some other solutions have been investigated like Multi-Layer IPsec. However, these solutions also require control case 1 and therefore the additional benefits of implementing such solutions are not significant. In the aeronautical communications scenario, due to the short-messages characteristics of the safety communication a gain in performance can be achieved when enhanced TCP, optimised for short messages is used. This is feasible to implement as it represents a control case 1, *and* recommendations by the satellite operator to the end user is possible. In the non-safety communication PEP could be implemented outside the IPsec channel between the airborne and ground APC IPsec GWs, but if the passenger's traffic is already IPsec-protected, the PEP will not bring any improvement.

3.2   Fragmentation

All VPN technologies add some overhead which might increase the packet size beyond the Path Maximum Transfer Unit (PMTU) value of one link in the end-to-end path. Therefore, after deploying a VPN the chances of fragmentation being required are higher. Fragmentation and reassembly are cumbersome. When reassembling, the node needs to wait for all fragments to be received. Fragmentation also adds some overhead, for each fragment requires its own IP header. Also, when happening at routers, they take more time and resources than simple forwarding, effectively increasing the computational load and the delay.

The situation is especially critical if fragmentation happens at the VPN gateway after packets are encapsulated. If that is the case, the receiving VPN gateway is forced to reassemble the packet before decapsulating it, rather than decapsulating each fragment and then forwarding them. This increases the computational load of a single node, the VPN gateway.

There are two proposed solutions. Whenever possible and feasible, the satellite link MTU should be configured to be greater than or equal to the packet sizes generated by both end host plus the VPN overhead. This assures that the packet does not require fragmentation at the satellite link. This solution requires control case 1, since it is required to know the MTU and the traffic characteristics of the end hosts.

Another solution is to use Path MTU Discovery [4]. That is, fragmentation is allowed only at the end hosts, but not at intermediate router.. This is the default behaviour for IPv6 and in IPv4 it is achieved by setting the Don't Fragment (DF) field in the IP header. Packets that exceed the MTU after being processed by the VPN should be dropped. The VPN GW then sends an ICMP message, indicating the path MTU (PMTU), back to the source so it can adapt the size of the packets to fit the tunnel MTU. When using IPv4, this solution requires control case 1.

In the aeronautical scenario, the safety communication uses IPv6 and PMTUD is used by default. In the non-safety communication, PMTUD is enforced at the APC IPsec gateway, by setting the DF bit of all egress IPv4 packets, so that it is dropped in case IPsec causes the packet size to be larger than the link MTU.

3.3   Overhead

VPNs add new headers, causing overhead, which increases the traffic load. However, the impact of this overhead will depend on the payload size. For smaller payloads (like VoIP) the relative increase will be important. For larger payloads the problem is not as bad, but it is still present.

To minimise the effect of such issue, it is proposed to apply header compression to to reduce the overhead. RObust Header Compression (ROHC) has been selected because it is currently the state of the art of header compression protocol that provides high compression efficiency and robustness [5].

This solution will be applied on the satellite link hop since it is the critical link on the path. ROHC can be used to compress ESP or AH headers of IPsec-protected packets. Even though ROHC can also be used to compress the TCP header of TLS-protected packets, the relative compression gain obtained is not significant, as the payload is usually already quite large (TCP will try to create packets with the size up to its maximum segment size (MSS), which takes into account the link MTU). For TLS it is recommended to use TLS-compression, which is part of the TLS specification [6]. ROHC is implemented in the satellite link for both VPN types, so it only requires control case 2.

3.4   Mobility

If at least one of the VPN peers is placed inside a mobile network, then some mobility related issues may appear. For a VPN connection, a handover becomes critical in case the IP address of one VPN peer changes since VPN security associations depend on IP addresses. This is the case for gateway, satellite and gap-filler (technology) handovers. If no mobility features are deployed, the VPN connection will become unusable and a new one has to be setup, requiring maybe a new user interaction for authentication (depends on the authentication mechanism), expensive calculations, and extra round-trips. This is especially critical for satellite links since they have a high round-trip time.

Some solutions have been proposed. One of them is MOBIKE which allows an IPsec security association to accept more than one IP address. The other, Mobile IP consists on giving the mobile node a static IP address (Home Address, HoA) additionally to the IP address given by the network it is plugged in (Care-of-Address, CoA). All packets sent to the HoA are routed through a static node called Home Agent, which knows at all times what CoA corresponds to each HoA. Mobile IP is extended from having a single mobile node to a mobile network with NEMO.

NEMO is the chosen solution for the aeronautical scenario. It is transparent to the end hosts and it is implemented in the satellite link, so it can be implemented even for control case 2. Therefore, it is useful for both safety and non-safety communications. MOBIKE's drawbacks, like being forced to start the communication on the airborne side, discard it as a viable option.

3.5   Quality of Service (QoS)

QoS provisions in the Internet are achieved through IntServ and DiffServ protocols [7] [8]. In both cases, the IP header is used to determine what treatment a datagram receives. Therefore, an IPsec VPNs in tunnel mode will be a problem because it is likely to encrypt the original IP header. A TLS VPNs will not be an issue as it does not protect the network layer.

To still being able to apply QoS at the satellite link (inside the IPsec tunnel), the required information should be present in the outer IP header. This is easily done when using DiffServ, as only the DS field is required, and it should be copied to the outer header according to the IPsec RFC [9]. IntServ access to other fields like source and destination addresses that are not available if the packets are tunnelled using IPsec. It requires control case 1 to assure that the IPsec implementation used at the gateway copies the value and so it can only be assured for the safety communications.

4   Conclusions

In this paper we explained the issues created when deploying a VPN in broadband satellite systems and more specifically, when the satellite links an airplane with the ground.

In case there is no control or influence over the elements in the communication, some of these issues remain untreatable, but for some others, a solution is still applicable. If all the elements are under control, there is at least one solution for each issue. While some solutions, like copying the DSCP value for the QoS issue eliminate the problem, some like header compression only minimise its effects.

Future work involves measuring the performance gains obtained from using the proposed solutions. For that, a testbed integrating the different elements of the aeronautical scenario and the solutions has been set up.

Acknowledgements. This has been conducted in the framework of ESA project "Security in broadband satellite systems for commercial and institutional applications". All fruitful discussions with the project partners and ESA are gratefully acknowledged. This paper does not implicitly represent the opinion of all project partners. The authors are solely responsible for it and do not represent the opinion of the ESA. ESA is not responsible for the use of any data in the paper.

References

1. Zhang, Y.: Multi-Layer Protection Scheme for IPsec. Internet Draft (October 1999)
2. SANDRA project, http://www.sandra.aero
3. Border, J., Kojo, M., Griner, J., Montenegro, G., Shelby, Z.: Performance Enhancing Proxies Intended to Mitigate Link-Related Degradations. RFC 3135 (June 2001)
4. Mogul, J., Deering, S.: Path MTU Discovery. RFC 1191 (November 1990)
5. Bormann, C., et al.: RObust Header Compression (ROHC): Framework and four profiles: RTP, UDP, ESP, and uncompressed. RFC 3095 (July 2001)
6. Dierks, T., Rescorla, E.: The Transport Layer Security (TLS) Protocol Version 1.2. RFC 5246 (August 2008)
7. Braden, R., Clark, D., Shenker, S.: Integrated Services in the Internet Architecture: an Overview. RFC 1633 (June 1994)
8. Blake, S., et al.: An architecture for Differentiated Services. RFC 2475 (December 1998)
9. Kent, S., Seo, K.: Security Architecture for the Internet Protocl. RFC 4301 (December 2005)

A New Dynamic Multilayer IPSec Protocol

Muhammad Nasir Mumtaz Bhutta and Haitham Cruickshank

CCSR, University of Surrey
Guildford, Surrey, United Kingdom
{m.bhutta,h.cruickshank}@surrey.ac.uk

Abstract. Performance Enhancing Proxies (PEPs) are used in satellite networks for better performance of the TCP/IP applications. Multi-layer IPSec (ML-IPSec) resolves the conflict between end-to-end security in standard IPSec and operation of PEPs. Previous Ml-IPSec solution has issues of limited application scope and increased complexity to implement and process the ML-IPSec protected data. This paper presents a new dynamic ML-IPSec protocol which addresses these issues. The paper also analyzes the protocol with reference to previous ML-IPSec protocol and presents the experiment performed to analyze the network performance while running IPSec and ML-IPSec.

Keywords: ML-IPSec, IPSec, PEP, TCP, Dynamic ML-IPSec.

1   Introduction

Multi-Layer IPSec (ML-IPSec) enhances the functionality of IPSec in order to solve the conflicts between IPSec and intermediate entities such as TCP and application layer PEPs. More information on Y. Zhang work on ML-IPSec can be obtained in [1], [2]. The earlier work on ML-IPSec, done by HRL Laboratories, was presented to IETF in many meetings and an internet-draft was written as well. IETF showed concern in three areas: 1) the idea presented by HRL Laboratories was only targeting very limited domain by fixing the zone map for the security association lifetime, 2) implementation complexity was increased and 3) it was required to show two more actual implementations of ML-IPSec. However, the problem of complexity of key management and security association setup for intermediate devices is also very complex and costly operation in terms of communication and it is not addressed very well. The HRL Laboratories suggested using "Internet Key Exchange (IKE v2)" for key setup. For large networks with large number of intermediate devices, using IKE v2 is not a good option. Also there are requirements for changing the databases of IPSec and IKE to make it compatible with ML-IPSec. The ML-IPSec analysis, design and IETF issues are discussed in detail by M.Bhutta and H.Cruickshank in [3], [4]. The issues are solved by our proposed novel, new dynamic ML-IPSec protocol. The paper also describes in detail the new proposed Dynamic ML-IPSec design and proof of study performed on SSFNet simulator.

P. Pillai, R. Shorey, and E. Ferro (Eds.): PSATS 2012, LNICST 52, pp. 119–129, 2013.

2   Previous Multilayer IPSec

First let us have an overview of ML-IPSec. The IP datagram is divided into portions. A portion under the same security protection scheme is called "Zone". A zone map is a mapping relationship from octets of the IP datagram to the associated zones for each octet. The zone boundaries must remain fixed within the lifetime of a security association otherwise it will be very difficult to do zone by zone decryption and authentication.

Security Association (SA) in IPSec defines the relationship between sender and receiver. The Composite Security Association (CSA) in ML-IPSec also includes the intermediate trusted nodes in addition to the sender and receiver. For each zone, there is an individual security association. Therefore, all security associations for all zones collectively form a CSA to cover the entire IP datagram. A CSA has two elements. The first element is zone map and second element is a zone list. Zone map shows the coverage of each zone in IP datagram and second element, zone list shows the list of SAs for each zone.

As Encapsulating Security Payload (ESP) in IPSec provides the maximum security features of IPSec protocol, so here our focus is on ESP only. The discussion on Authentication Header (AH) is out of scope of this paper.

The ESP payload data field in ML-IPSec is divided into multiple pieces depending upon the number of zones. The payload data for each zone collectively along with padding, padding length and next header field is referred to as cipher text block of the zone. In ML-IPSec, different IP datagram parts can be encrypted using different keys for different zones. The ESP authentication data field is also variable in length and contains multiple ICVs which are calculated for different zones and the size of them is dependent on the algorithms being used for integrity. More information on previous ML-IPsec can be found in [1], [2].

3   Issues in Previous ML-IPSec

As notified by IETF, the application scope and increased implementation complexity are main issues in previous ML-IPSec [5, HRL Laboratory report]. Also, key management for ML-IPSec is a very complex and big concern to make ML-IPSec enable to provide security services. Y. Zhang proposed to use IKEv2 to establish the security associations between intermediate communicating parties but, using IKEv2 will not scale well and complexity will also increase as network will grow. However, the main focus here is to address the limited application scope of the previous ML-IPSec protocol. The key management complexity is out of the scope of this paper.

The main reason for limited application scope of previous ML-IPSec protocol is the way how zone map is established. As described earlier in section II, zone map defines the coverage of each zone in IP datagram. The zone map is part of composite security association (CSA) and is established between communicating parties when security association (SA) is established. The zone map must remain constant for the duration of established security association (SA) life time. By making zone map constant, restricts many applications to use ML-IPSec like HTTP in which request/response size is variable due to appending cookies etc or any application appending extra headers will also not work for constant zone map. This paper presents a new dynamic ML-IPSec protocol which increases the scope of application and address the constant zone map problem.

4   New Dynamic ML-IPSec Protocol

The proposed new dynamic ML-IPSec protocol increases the application support by allowing breaking down the IP datagram into zones as per requirement. Before describing the details of new ML-IPSec protocol, we summaries the design considerations here. We propose that zone map should not be part of CSA; instead the zone information is embedded in the ESP header. CSA will remain same except the zone map information. It will contain the zone list and all designated and non-designated security association parameters will also remain as part of CSA and will be described in the same way. The zone information about the IP datagram will go as part of ESP.

The inbound, outbound processing on participating nodes, ESP and AH packets parsing and security processing on ESP and AH are affected by making zone information part of ESP and AH header. The paper only focuses on ESP but, the basic logic and processing with respect to zones will be same for AH as well. All these processing procedures and design details are described in this section.

4.1   Zones and Zone Map

A zone in new ML-IPSec is a continuous block (portion) of IP datagram. The reason to identify a zone as a continuous block is due to the reason as it reduces the complexity to process a part of IP datagram. The detailed discussion on zone manipulations is in the coming sub-section, Security Protocols, where we discuss all the processing details.

A zone map in previous ML-IPSec was described as bit-by-bit mapping of IP datagram into zones. However, in new ML-IPSec protocol zone map also contains information about zones but, zone map is combinations of zone pointers. A zone pointer points to the starting bit location in ESP header when IP datagram is encapsulated after encryption. Further details about zone map are described in sub-section, Security Protocols, along with parsing and processing details of ESP header and IP datagram.

4.2   Composite Security Association (CSA)

CSA in previous ML-IPSec consists of two elements, zone list and zone map. Zone list is a list of all security associations associated with each zone and zone map defines the IP datagram bits associated with each zone. However, CSA only consists of zone list without zone map as zone map information becomes part of ESP header. Except this change, CSA remains unchanged including the definitions of designated and non-designated parameters.

4.3   Security Protocol

As described in section II.C, the paper focuses only to discuss security protocols with respect to ESP only.

The figure 5 shows the new ESP header. The ESP header is also the same as IPSec and previous ML-IPSec ESP header with some changes.

4.3.1 ESP Header Design

The ESP header contains SPI and sequence number as in IPSec and previous ML-IPSec. After sequence number field, the ESP header contains a 4 bit field called "Total Zones". This field identifies the total number of zones of each IP datagram under process. The IP datagram can be broken into as many zones as required by application up to a maximum of 15 zones.

After "Total Zones" field, there are variable numbers of pointers fields depending upon the number of zones. Each pointer field consists of 11 bits to support the IP MTU of 1400 bytes and points to the starting position of each zone inside ESP payload. After adding all the pointer fields, the necessary padding is done to keep ESP header word size constant of 32 bits. The "Pad Length" 5 bits field describes the total number of padding bits added in the ESP header for pointers.

The ESP payload data in new ML-IPSec protocol consists of multiple pieces depending upon the number zone. The data in each zone is encrypted after adding any necessary padding and then ICV is calculated on the resultant cipher data for that zone. The cipher data and ICV for a specific zone is combined and put in the ESP payload data at specific position. The resultant encrypted data and ICV of different zones are encapsulated in the ESP payload in the order consistent with IP datagram data.

The figure 5 shows the new ESP header for n number of zones.

4.3.2 ESP Header Processing

The ESP header processing at participating nodes is in consistence with previous ML-IPSec and IPSec protocols. The processing steps described here are followed in outbound and inbound processing performed in participating entities in the communication. The processing steps are as follows:

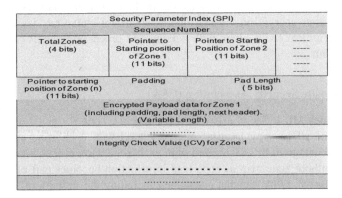

Fig. 1. ESP Header for New ML-IPSec Protocol

- For encryption, the payload data part for each zone is appended with necessary padding and pad length depending upon the algorithm and then encrypted. However, the designated zone also contains the "Next Header" field as in previous ML-IPSec. For decryption operation, the decryption is done depending upon the algorithm selected and original data is received after truncating the padding and pad length fields. The operation steps performed for encryption/decryption are in consistence with IPSec and ML-IPSec.
- The ICV calculation steps in new ML-IPSec are same as in IPSec and previous ML-IPSec depending upon the selected algorithm. However, the way ICV for each zone is encapsulated in ESP header is different from previous ML-IPSec. In previous ML-IPSec ICVs for all zones were combined together and then appended at the end of combined encrypted data of all zones but, in new ML-IPSec ICV for each zone is appended at the end of encrypted data for that specific zone to make it easy for processing and less complex. As ICV is always calculated to fixed size depending upon the algorithm selected, it can be easily extracted from the end of zone encrypted data at receiving end to verify the integrity.

4.4   Inbound, Outbound Processing

The outbound and inbound processing can be referred as processing at sender end and receiver processing at end respectively. In intermediate nodes, partial processing is done on IP datagram accessible part; however the steps remain same with respect to inbound and outbound processing.

4.5   Outbound Processing in ML-IPSec

In new ML-IPSec protocol the outbound processing is done in the same way as described in section II.D with the following exceptions:

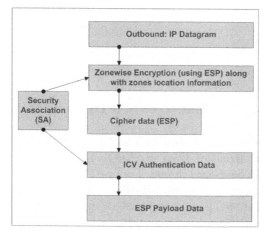

Fig. 2. Example of Outbound Processing

- There is no pre-defined constant zone map which is consulted to perform security operations on each selected zone. The zones are generated dynamically and are processed for encryption and integrity check value calculation. However, the sequence of performing security operations remains same.

The processing steps for outbound processing are shown in figure 6.

4.6   Inbound Processing in ML-IPSec

The inbound processing in ML-IPSec is reverse of the outbound processing. The processing steps for inbound processing are shown in 7:

5   Security Services Offered by New ML-IPSec Protocol

The new ML-IPSec offers same security services as offered by IPSec and ML-IPSec including origin authentication, connectionless integrity, optional partial sequence integrity, data confidentiality and limited traffic flow confidentiality with the help of ESP.

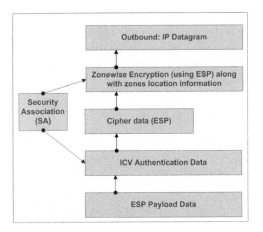

Fig. 3. Example of Inbound Processing

6   Performance Evaluation of New Dynamic ML-IPSec Protocol

To evaluate the performance of new proposed dynamic ML-IPSec, we selected to modify the implementation of IPSec by NIST. The simulator used by NIST was SSF/SSFNet and was developed in Java. For proof of study and analyze the network performance while running Dynamic ML-IPSec, we have modified the NIST IPSec implementation according to our proposed design. Following are the details of experiment performed.

6.1   Experiment Environment

A network of asymmetric kind where one part of network only contained clients and one part of network only contained servers were arranged in dumbbell topology as shown in figure 8. The network consisted of a pair of security gateways and one or more hosts behind each gateway (connected with a LAN). Each gateway provides secure VPN services using Dynamic ML-IPSec and IPSec for the local hosts and acts as the SA initiator or the SA responder. All the experiments were performed in tunnel mode where a security policy controls the packets to be processed. For our experiment, we configured the host behind security gateway to create a security association with each other node.

In the experiment, we used manual key management. So at start of experiment, a SA was established for IKE and for IPSec/ML-IPSec. In ML-IPSec experiment, the SAs were different for different zones. The life time of SA was more than the experiment time, so that no re-keying should be required for life time of experiment.

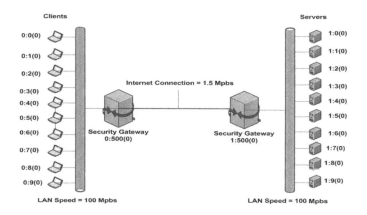

Fig. 4. A sample of network configuration used for experiment

6.2   Traffic Models

In the simulation, different experiments were performed for file transfer between TCP clients and TCP servers. A client connects to a randomly chosen server and requests server to transfer a file of fixed size for the selected duration of experiment. A TCP-based application continuously generates fixed number of files traffic for each session for the duration of the experiment. The TCP clients were waiting for a random time to send the next request after completion of the session.

6.3   Performance Measures

The purpose of the study was proof of concept of dynamic ML-IPSec that it functions according to the policy and we achieved similar behavior between IPSec and

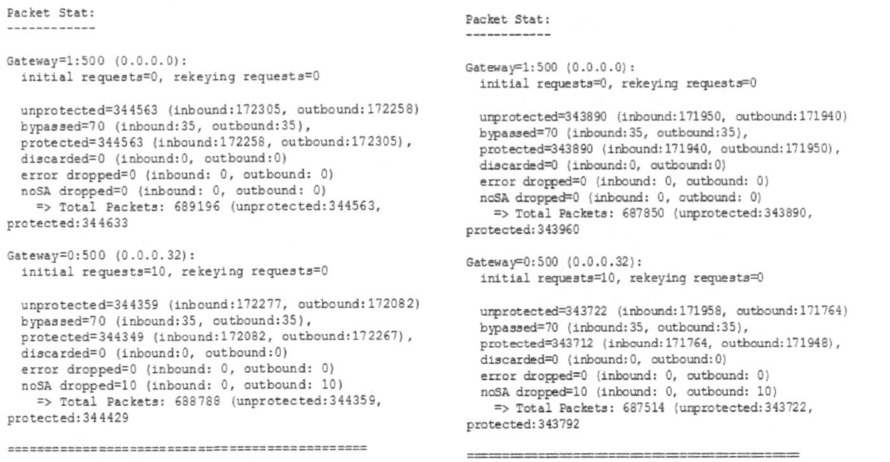

```
Packet Stat:
------------

Gateway=1:500 (0.0.0.0):
  initial requests=0, rekeying requests=0

  unprotected=344563 (inbound:172305, outbound:172258)
  bypassed=70 (inbound:35, outbound:35),
  protected=344563 (inbound:172258, outbound:172305),
  discarded=0 (inbound:0, outbound:0)
  error dropped=0 (inbound: 0, outbound: 0)
  noSA dropped=0 (inbound: 0, outbound: 0)
    => Total Packets: 689196 (unprotected:344563,
protected:344633

Gateway=0:500 (0.0.0.32):
  initial requests=10, rekeying requests=0

  unprotected=344359 (inbound:172277, outbound:172082)
  bypassed=70 (inbound:35, outbound:35),
  protected=344349 (inbound:172082, outbound:172267),
  discarded=0 (inbound:0, outbound:0)
  error dropped=0 (inbound: 0, outbound: 0)
  noSA dropped=10 (inbound: 0, outbound: 10)
    => Total Packets: 688788 (unprotected:344359,
protected:344429

=========================================
```

Packet Statistics for IPSec

```
Packet Stat:
------------

Gateway=1:500 (0.0.0.0):
  initial requests=0, rekeying requests=0

  unprotected=343890 (inbound:171950, outbound:171940)
  bypassed=70 (inbound:35, outbound:35),
  protected=343890 (inbound:171940, outbound:171950),
  discarded=0 (inbound:0, outbound:0)
  error dropped=0 (inbound: 0, outbound: 0)
  noSA dropped=0 (inbound: 0, outbound: 0)
    => Total Packets: 687850 (unprotected:343890,
protected:343960

Gateway=0:500 (0.0.0.32):
  initial requests=10, rekeying requests=0

  unprotected=343722 (inbound:171958, outbound:171764)
  bypassed=70 (inbound:35, outbound:35),
  protected=343712 (inbound:171764, outbound:171948),
  discarded=0 (inbound:0, outbound:0)
  error dropped=0 (inbound: 0, outbound: 0)
  noSA dropped=10 (inbound: 0, outbound: 10)
    => Total Packets: 687514 (unprotected:343722,
protected:343792

=========================================
```

Packet Statistics for ML-IPSec (with 1 zone)

Fig. 5. Example of Inbound Processing

ML-IPSec for same set of security related experiments. We also tried to compare the network performance while running IPSec and ML-IPSec for different traffic loads and different network bandwidth configurations.

The packet statistics shown in figure 9 are counted according to the security policy enforcement and selected network conditions. We counted the protected/unprotected/bypassed/error packets for both inbound and outbound independently within a security gateway.

For the application layer, we gathered the session throughput for different network conditions. We run the simulation in tunnel mode for fixed time. The internet connection between securities gateways were configured for different bandwidth values from 1 Mbps to 10 Mbps to see the effects of network bandwidth with respect to traffic load. To change the traffic load, we changed the file sizes from 1 Mbytes to 20 Mbytes. The security attributes like 3DES_CBC, AES or HMAC_SHA1 etc, can change the overall performance under specific environment. We considered the affects of our selected security schemes to play a role in our analysis. The cryptographic figures used for our experiment are given below and are in consistence with NIST IPSec experiment.

Table 1. Cryptographic Figures used for experiment

Algorithms	Block Size	Key Size
3_ DES_CBC	8 bytes	24 bytes
HMAC_SHA1	64 bytes	20 bytes

As shown in figures 10 and 11, overall throughput of network becomes stable once the network can handle all the transmitted data efficiently than the speed data is transmitted on the network. There remains unnoticeable difference between the performance while running IPSec and ML-IPSec with different number of zones. However, when network speed is less than the speed at which processor is transmitting the data on the network; we have observed that configuration where less processing is involved gives better performance as compared to configuration where high processing is required. Hence in low bandwidth network, obviously IPSec gives better performance as compared to ML-IPSec. However, this difference is not very high. The below graph and data table shows the average throughput obtained in the analysis.

Table 2. Network Throughput for file size of 20MBytes

File Size (MB)	Bandwidth (Mbps)	Average Throughput			
		IPSec	ML-IPSec (1 Zone)	ML-IPSec (2 Zones)	ML-IPSec (3 Zones)
20	1	92.842	92.669	91.906	91.822
20	2	185.095	184.74	183.235	183.07
20	3	251.796	251.793	251.776	251.776
20	4	252.02	252.02	252.01	252.006
20	5	252.13	252.13	252.13	252.13
20	6	252.23	252.23	252.22	252.22
20	7	252.286	252.286	252.286	252.283
20	8	252.336	252.333	252.33	252.33
20	9	252.336	252.336	252.33	252.33
20	10	252.403	252.4	252.396	252.396

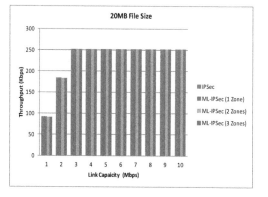

Fig. 6. Network Throughput for file size of 20MBytes

The figure also shows that as file size increases, the TCP application gives better utilize the network bandwidth as compared to low file sizes. The low file size degrades the network performance to very small extent.

6.4   Limitations of Implementation

The analysis was performed using SSF/SSFNet simulator and NIST IPsec implementation. However, for our analysis there are some limitations inherited from NIST IPSec implementation and SSF/SSFNet implementation which are given below:

- No actual implementation of cryptographic algorithms, keys and cryptographic operations is done. However, the security processing behavior was simulated alongside processing block size and processing time.
- The header sizes and data sizes may be different from actual implementations for different constraints of Java language.

However, the overall behavior of IPSec, ML-IPSec will not be some much different from real implementation.

7   Conclusion

The ML-IPSec can solve the interworking issues between intermediate devices such as PEPs and IPSec. ML-IPSec enables PEPs to access a limited portion of IP datagram for their proper functioning while end-to-end data confidentiality is preserved by ML-IPSec. However, there are some issues in previous ML-IPSec solution like limited application domain, which can be resolved by new dynamic ML-IPSec by making application more flexible to break IP datagram into different zones. The new dynamic ML-IPSec also improves the efficiency and reduces complexity to encapsulate the zones information into ESP payload.

The paper presented new dynamic ML-IPSec with detailed description of its design and processing. The paper has also performed an analysis on new dynamic ML-IPSec in comparison with IPSec and previous ML-IPSec where appropriate. The paper has shown some results of our analysis for security policy enforcement and network performance evaluation with different network bandwidth and traffic load. It is observed that ML-IPSec gives almost same performance as IPSec performs when network bandwidth is more than 3 Mbps. However, when network bandwidth is low, then there is small performance reduction as compared to IPSec.

References

[1] Zhang, Y.: A Multilayer IP Security Protocol for TCP Performance Enhancement in Wireless Networks. IEEE Journals on Selected Areas in Communicaitons 22(4) (May 2004)
[2] Zhang, Y., Singh, B.: A multi-layer IPsec protocol. In: Proc. Usenix Security Symp., pp. 213–228 (August 2000)
[3] Cruickshank, H., Bhutta, M.N.M., Ashworth, J., Moseley, M.: Interworking between Satellite Performance Enhancing Proxies and Multilayer IPSec (ML-IPSec). In: 16th KA and Broadband Communications 2010, Milan, Italy (2010)

[4] Bhutta, M.N.M., Haitham, Ashworth, J., Moseley, M.: Multilayer IPSec (ML-IPSec) Design for Transport and Application Layer Satellite Performance Enhancing Proxies. In: 28th AIAA International Communications Satellite Systems, AIAA/ICSSC, Anaheim, California (2010)

[5] Zhang, Y.: HRL Laboratories Report, Multi-layer Internet Security for Satellite and Wireless Networks (December 1999)

[6] Border, J., et al.: Performance Enhancing Proxies Intended to Mitigate Link-Related Degradations. IETF RFC 3135 (June 2001)

[7] Cruickshank, H.: Technical Report on Performance Enhancing Proxies (PEPs) for the European ETSI Broadband Satellite Multimedia (BSM) working group. ETSI Report TR 102 676 (September 2009), http://portal.etsi.org

[8] Gomez, C., et al.: Web browsing optimization over 2.5G and 3G: end-to-end mechanisms vs. usage of performance enhancing proxies. Wireless Communications and Mobile Computing 8, 213–230 (2008)

[9] Kent, S., Seo, K.: BBN Technologies, Security Architecture for Internet Protocol. RFC 4301 (December 2005)

[10] Kent, S.: BBN Technologies, IP Authentication Header (AH). RFC 4302 (December 2005)

[11] Kent, S.: BBN Technologies, IP Encapsulating Security Payload (ESP). RFC 4303 (December 2005)

[12] Kaufman, C.: Microsoft, Internet Key Exchange (IKEv2) Protocol. RFC 4306 (December 2005)

[13] Obanaik, V.: Secure performance enhancing proxy: To ensure end-to-end security and enhance TCP performance over IPv6 wireless networks. Elsevier Computer Networks 50, 2225–2238 (2006)

[14] Bellovin, S.: Probable plaintext cryptanalysis of the IPSecurity protocols. In: Proceedings of the Symposium on Network and Distributed System Security (February 1997)

[15] Dierks, T., et al.: The TLS Protocol Version 1.2, RFC 5246 (August 2008)

[16] Sing, J., Soh, B.: A Critical Analysis of Multi-layer IP Security Protocol. In: Third International Conference on Information Technology and Applications, ICITA 2005 (2005)

[17] Annoni, M., Boiero, G., Salis, N., Cruickshank, H.S., Howarth, M.P., Sun, Z.: Interworking between multi-layer IPSEC and Secure multicast services over GEO satellites. Eur. Cooperation in the Field of Sci. Tech. Res., Tech. Rep. COST 272 TD-02–016 (2002)

[18] Annoni, M., Boiero, G., Salis, N.: Security issues in the BRAHMS system. In: Proc. Ist MobileWireless Telecommunications Summit 2002 (June 2002)

[19] Baugher, M., et al.: Multicast Security (MSEC) Group Key Management Architecture. IETF RFC 4046 (April 2005)

[20] Cruickshank, H.: Technical Specifications for satellite networks multicast security architecture and key management for the European ETSI Broadband Satellite Multimedia (BSM) working group. ETSI Specifications. ETSI TS 102 466 (December 2006), http://portal.etsi.org

[21] Wallner, D., et al.: Key Management for Multicast: Issues and Architectures. IETF RFC 2627 (June 1999)

[22] Sirsuresh, P., et al.: Middlebox Communication Architecture and Framework. IETF RFC 3303 (August 2002)

Efficient Integration of Satellites
in Collaborative Network Structures

Bernd Klasen

SES ASTRA TechCom, L-6815 Betzdorf, Luxembourg
bernd.klasen@ses.com

Abstract. Technological advances are enabling the reception of satellite broadcast even on mobile devices. However, mobile device usage is highly individual and exhibits an on-demand characteristic which rather suggests unicast. In order to avoid a wastage of the strengths of satellite broadcasts, additional effort is needed. This paper presents an approach to integrate satellites into heterogeneous collaborative networks for provision of IP services and broadband Internet access. Thus it supports both real time multimedia streaming and multimedia on demand.

Keywords: IP over satellite, network convergence, heterogeneous networks, content distribution.

1   Introduction

Since the establishment of the Internet and basically with the launch of the World Wide Web (WWW), human life considerably changed. It is affecting the way we work as well as the way we spend our leisure time. We look for Information, buy and sell, play games and communicate with other people in forums, chats and social networking platforms. The importance of online activities and the time we donate to them increases with the enhancing availability and growing bandwidth of the Internet accesses. At the same time we accustom ourselves to the way the Internet provides us with things we are looking for – which is immediate, ubiquitous and on–demand. Everything is just a few clicks away and waiting times of several seconds are considered as long in this context.

Due to the success of smart–phones and other mobile devices that encourage us to an ubiquitous Internet usage, this effect has even been further amplified. Our desire to exchange information, to participate in social networks, to use the latest software – the word apps is often used in this context – and to access videos and other multimedia files always and everywhere leads to constantly increasing network traffic and bandwidth demands. Permanent upgrades and investments on the infrastructure are needed in order to preserve connection speed and service availability under this growing load. Considering this situation, being now able to receive satellite broadcasts on mobile devices might seem to be the solution to all present problems. However, satellites – and broadcast networks in general – are inherently designed to serve a large number of people with the same content at the same time, which is in contrast to the highly

P. Pillai, R. Shorey, and E. Ferro (Eds.): PSATS 2012, LNICST 52, pp. 130–137, 2013.

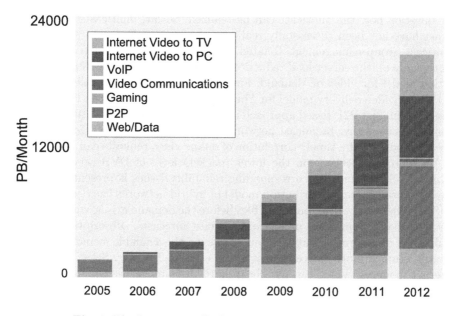

Fig. 1. The Internet traffic forecast. Source: www.cisco.com

individual behavior described above. In case the satellite bandwidth must be split among thousands or even millions of users for individual transfers, the possible benefits are extinguished. As a solution to this problem, this paper presents an approach for efficient satellite integration into mobile networks that uses resource request aggregation. Files that are downloaded by a large number of users nearly simultaneously are being broadcasted, while less popular data is send via unicast. The following sections will show that this approach not only will cut the peak loads caused by "hot" or "hype" content but also will allow better QoS for the most traffic intense type of online content. The latter, according to studies presented in [10] and to the Cisco Internet traffic forecast shown in figure 1, are video files. It is worth noting that although P2P traffic seems to cause at least the same amount of traffic, we have to recall that a large fraction of files exchanged via P2P consists of multimedia content. Due to the high relevance of video files for the overall Internet traffic, the proposed network model is motivated and described in the scope of video files. However, it can be also applied to other types of content with similar characteristics regarding the access patterns.

2 Related Work

The potential for traffic savings by aggregating viewers and then using an Internet multicast has been analyzed in [1]. While this study shows the applicability of request aggregation in an video on demand environment, it does not answer the question how the multicast can be realized. So far, multicasts over the Internet have not been successfully realized in a large scale due to the effort for multicast–group management. Related work has also been presented by [5] who concentrate on the server load reduction potential when using a P2P–like distribution model for video on demand. Further the authors of [4] analyzed also an P2P based video delivery model for YouTube which had a focus on the reception side. Also in [2] a P2P based approach is analyzed. None of these studies utilized a satellite or another broadcast network for increased efficiency. However, their research shows that a timely correlation of online video requests can be observed.

An approach focusing on the lower (packet) level of DVB service delivery to mobile clients and the corresponding reliability issues is presented in [7]. A satellite based content distribution model in hybrid networks has been presented in [8]. It relies on the distribution of files before the demand arises, which becomes possible due to popularity predictions, request forecasts, subscriptions and the availability of large caches at the users. However, the available memory on mobile devices is much more scarce than it is the case on TV–sets, Set–Top–Boxes or Internet modems, which can use cheap hard disks as caches that provide high capacities. Thus for mobile networks, a different approach must be taken. Let us assume that we have enough storage to completely cache the video currently played which is reasonable considering current standards.

3   Content Distribution Model

Since online video on demand (VoD) allows users to start playback of a video at an arbitrary moment in time, usually it is sent to the customer by the time playback is started as unicast, which is natural choice considering the individual nature of this service. But in case thousand people start watching at the same moment, the whole data is sent one thousand times. Besides of the massive amount of traffic this generates, there is another problem: If the average bandwidth of the user does not at least match the video's encoding bitrate, a fluent playback is not possible or he has to wait a long time before enough data is buffered. Considering the numerous rural areas where UMTS is still not available – not even to talk about LTE – this is an important aspect considering mobile video reception.

Before we start with the details of the model, let us be reminded on an important property that is essential for the applicability of the described approach. This property is the power law distribution of online video popularity which is confirmed by [6]. Similar results are presented by the authors of [3] who state that 10% of all available videos cause 90% of all web traffic. While many studies pay special attention on the long tail of that distribution, the model described in this paper concentrates on the opposite, on the most popular files. It has further been shown in [9] that there are online videos which exhibit heavy bursts in their request patterns. The fact that there exist extremely popular videos which at the same time show a strong timely correlation in their view statistics ensures that there are files which can efficiently distributed by using the model presented in this paper. As already mentioned, the latter is roughly based on the rule that files with many concurrent requests are broadcasted while the others are transmitted via the terrestrial network as unicast. In detail, the model works as follows:

Like it is common for peer-to-peer file exchange protocols, we subdivide each file into several pieces of size S_P. We choose the piece size accordingly to the available mobile network bandwidth B_M so that one piece can be submitted per second. This also means that this is the maximum encoding bitrate that can be used for the video in order to allow fluent playback. Later we will also see simulation results where a higher video encoding bitrate is used which means that these videos could not have been viewed instantaneously in full quality when only the mobile network with EDGE bandwidth was used for delivery. Regardless of a specific encoding bitrate, given a size S_{F_i} for a file F_i, this means that F_i is split into

$$|P_{F_i}| := \frac{S_{F_i}}{S_P} \tag{1}$$

pieces. Let further the available satellite bandwidth be B_S. Whenever the number of active downloaders exceeds a defined broadcast threshold BT, a broadcast is scheduled. The full satellite bandwidth is utilized then, which means that we are able to transmit multiple file pieces within one second. The ratio r between the files transmitted per second via mobile network and via broadcast depends on the available mobile network bandwidth. Thus it is

$$r = \frac{B_S}{B_M} \tag{2}$$

Since we assume that all pieces are received sequentially via mobile network, when BT is reached we check which is the piece with the highest sequence number $piece_{max}(F_i)$ that at least one on the active downloaders already holds. Then during the next step respectively the next second, r pieces are submitted at once via broadcast. In case that

$$piece_{max}(F_i) + r > |C_{F_i}| \tag{3}$$

we change the start sequence number for the broadcasted pieces to $|C_{F_i}| - r$ in order to utilize the full available bandwidth. This means in that special case some pieces are broadcasted even if BT is not reached for them, but it further increases the download speed for the remaining clients. Also, all clients that would not have reached the sequence number $piece_{max}(F_i)$ after the broadcast, the next piece in line is transmitted via unicast just as it would have been the case without a broadcast. Two other parameters we need are the maximum number of clients that can start watching a video per second (CpS) and the probability P that they will. The following evaluation of this model will show the possible traffic savings.

4 Evaluation

The first important question concerning the evaluation of the proposed model is the choice of the parameters. Regarding the mobile network bandwidth, choose EDGE with 236 Kbit/sec as the lowest common denominator since it is still the maximum available network bandwidth for several regions. Thus let

$$S_P = 236kbit \tag{4a}$$
$$B_M = 236kbit/sec \tag{4b}$$
$$B_S = 36,000kbit/sec \tag{4c}$$

A very critical parameter is the number of users that might request a certain file per step respectively per second. Here we use results of former studies that analyzed video request rates on YouTube [9]. According to those, there are videos which exhibit an average of 715.5 requests per minute during the first three days after they have been uploaded.

This means we have approximately 12 requests per second. Since the data is relatively coarse grained, it is impossible to distinguish how these requests are distributed within a period shorter than one hour. Thus we assume a Poisson distribution and let the

$$CpS = 7,155 \tag{5a}$$
$$P = 0.0017 \tag{5b}$$

This corresponds to the probability for a client to request the file within 10 minutes. Further we use the average video duration of 342 seconds, which relies on the results for very popular videos from the YouTube study mentioned above. Considering the now commonly used average bitrate of $1,057Kbps$ for

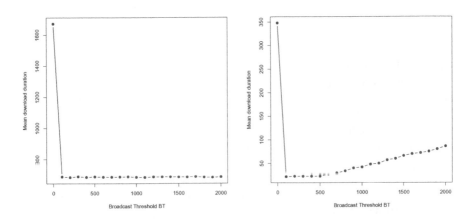

Fig. 2. Encoding bitrate = 1057 Kbps   **Fig. 3.** Encoding bitrate = 236 Kbps

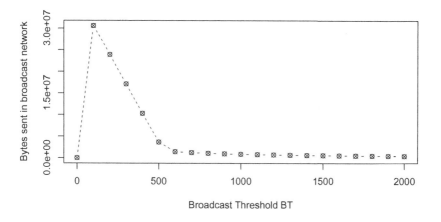

Fig. 4. kByte broadcasted

video with resolution 480p – which is not the highest quality but a common resolution we find on screens of mobile devices – and a audio bitrate of $96Kbps$ it results in an average video file size of $\approx 48.16MByte$. We also use a lower bitrate of $236Kbps$ with a corresponding filesize of 10.01 MB which allows a fluent playback without an excessively long buffering period when we have no broadcasts. Figure 2 and 3 show the download durations for varying values of BT for both encoding bitrates, while $BT = 0$ means that broadcasts are disabled. For the higher bitrate (figure 2) we have an average bandwidth saving in the mobile network of 58%, while it is more than 86% in case of the lower bitrate (figure 3). These are average values, and what we can clearly see in figure 3 is that the time a client needs to complete the download immediately increases

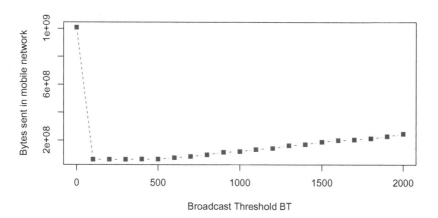

Fig. 5. kByte unicasted

with growing values of BT. This is what one expects considering the smaller filesize. In general it can be stated that the larger a file – and thus the longer the expected download time – the higher we can choose BT while still achieving a high mobile network traffic reduction. Figure 4 shows the broadcast bandwidth, figure 5 the mobile bandwidth needed under changing values of BT. Again, for $BT = 0$ the broadcast is disabled.

5   Conclusion

This paper presents a content distribution model that uses satellite broadcasts in a collaborative manner together with a terrestrial unicast network in order to reduce network traffic. It has been shown that the unicast traffic can be reduced by more than about 86%. At the same time, only relatively little satellite bandwidth is needed. This approach requires that some authority – for example the mobile or satellite network provider – keeps track of the number of current requests for a certain file and decides on the delivery channel accordingly. Since this violates the net–neutrality, it is important to implement the described approach in a way that users are aware of this and use it voluntarily. The reason why people should freely decide to do this is a fluent video playback even under high network load and shorter download times for videos. The latter can be very important if people spend also time in environments where no network access is possible.

Acknowledgment. This research is supported by the *National Research Fund* (FNR) of Luxembourg.

References

1. Aggarwal, V., Caldebank, R., Gopalakrishnan, V., Jana, R., Ramakrishnan, K.K., Yu, F.: The effectiveness of intelligent scheduling for multicast video-on-demand. In: Proceedings of the Seventeen ACM International Conference on Multimedia - MM 2009, p. 421 (2009)
2. Carlsson, N., Eager, D., Mahanti, A.: Peer-assisted on-demand video streaming with selfish peers. In: Networking 2009, pp. 586–599 (May 2009)
3. Cha, M., Kwak, H., Rodriguez, P., Ahn, Y.-Y., Moon, S.: I tube, you tube, everybody tubes: analyzing the world's largest user generated content video system. In: IMC 2007: Proceedings of the 7th ACM SIGCOMM Conference on Internet Measurement, pp. 1–14. ACM, New York (2007)
4. Cheng, X., Liu, J., Wang, H.: Accelerating YouTube with video correlation. In: Proceedings of the First SIGMM Workshop on Social Media - WSM 2009, p. 49 (2009)
5. Garbacki, P., Epema, D.H.J., Pouwelse, J., Van Steen, M.: Offloading servers with collaborative video on demand. In: Proceedings of the 7th International Conference on Peer-to-Peer Systems, p. 6. USENIX Association (2008)
6. Gill, P., Arlitt, M., Li, Z., Mahanti, A.: Youtube traffic characterization: a view from the edge. In: Proceedings of the 7th ACM SIGCOMM Conference on Internet Measurement, pp. 15–28. ACM (2007)

7. Gotta, A., Barsocchi, P.: Experimental video broadcasting in DVB-RCS/S2 with land mobile satellite channel: a reliability issue. In: Fourth International Workshop on Satellite and Space Communications 2008 (2008)
8. Klasen, B.: Efficient Content Distribution in Social Aware Hybrid Networks. Journal of Computational Science (2011)
9. Klasen, B.: Social, fast, efficient: Content distribution in hybrid networks. In: 2011 IEEE Symposium on Computers and Communications, ISCC, pp. 61–67. IEEE, Kerkyra (2011)
10. Maier, G., Feldmann, A., Paxson, V., Allman, M.: On dominant characteristics of residential broadband internet traffic. In: Proceedings of the 9th ACM SIGCOMM Conference on Internet Measurement Conference, vol. 9, pp. 90–102. Citeseer, New York (2009)

Physical Layer Representation for Satellite Network Emulator

Santiago Peña Luque, Emmanuel Dubois, Fabrice Arnal,
Patrick Gélard, and Cedric Baudoin

[1] Centre National d'Etudes Spatiales (CNES),18 av. Edouard Belin, 31401 Toulouse, France
[2] Thales Alenia Space, 26 Avenue JF Champollion, BP 33787, 31037 Toulouse, France
{santiago.penaluque,emmanuel.dubois,patrick.gelard}@cnes.fr,
{cedric.baudoin,fabrice.arnal}@thalesaleniaspace.com

Abstract. In general, emulation of satellite networks has been focused on representing the features of the access network, characterised by large complexity. Physical layer impact has usually been neglected or reduced to static and limited models. The aim of this work is to reformulate a new model of physical layer oriented to network emulation, adapted to multiple satellite systems and configurations. The main achievement of this work is the design of a simple method of analysis of end-to-end communication links, that can rely on separated uplink and downlink attenuation channels generated offline or in runtime. The model has been integrated in the open source platform PLATINE[1], showing an easy integration to distributed machine emulation and reproducing local propagation conditions for each ground terminal.

Keywords: satellite, emulation, network, physical layer, DVB-S2, DVB-RCS, BER, ACM, delay, C/N.

1 Introduction

In the context of networks based on bidirectional satellite systems, gateway stations are connected via satellite links to an important number of terminals, sharing limited bandwidth. The geographic distribution of ground terminals may result in very different atmospheric conditions and interference levels for each satellite link. Also, due to different uplink and downlink frequencies, the propagation conditions may differ slightly in each direction of the communications (forward/return link). Consequently, the resulting network is composed of communication segments whose performances vary throughout space and time (similar to the local conditions of each ground terminal), defining an important aspect of satellite network architectures, as presented in figure 1. Rain especially may strongly affect and collapse a communication link between the satellite and a ground station. Network resources should also implement adaptability to these changing scenarios. For example, in the case of DVB-S2/RCS

---

[1] The license of this project has recently changed to free software license GPL. Under the new conditions, the name of PLATINE will be abandoned and a new name for the project will be applied in the future.

P. Pillai, R. Shorey, and E. Ferro (Eds.): PSATS 2012, LNICST 52, pp. 138–147, 2013.
© Institute for Computer Sciences, Social Informatics and Telecommunications Engineering 2013

architecture, featured with Adaptive Coding and Modulation (ACM) in the forward link, the Network Control Centre (NCC) shall define how to broadcast content with the most efficient MODCOD schemes at all times, and considering the different (C/N_0) reception conditions of each user [1].

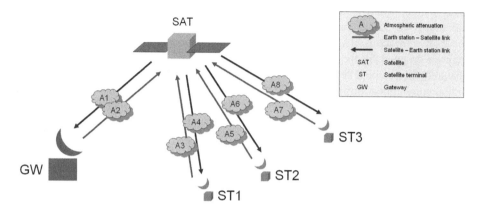

Fig. 1. Attenuation in different links in DVB-S2/RCS based in a unique satellite architecture

In this context, we want to build a model that reflects the main impacts of the physical layer on network communications, such as error introduction and insertion of delays, while still keeping the perspective of network emulation. In other words, the main objective of this paper is to present the base for a physical layer model, simple enough to be efficient in terms of computation, and representative enough to test network performance and capabilities in the context of satellite communications. Additionally, it will be required that the model present a flexible logical structure and several configurations, from extremely simple to more complex models. Network access emulation is already a complex system; the objective is to provide a simple and configurable model to evaluate more realistic results regarding the physical layer impact on satellite networks.

The second section of this paper recalls the principles of analysis of RF communications for both single and multiple links. The third section presents the architecture of PLATINE and the previous context of physical layer representation over satellite networks. The fourth section describes an implementation proposal for the analysis of the Physical Layer impact on bidirectional satellite networks. Final conclusions for this study are presented in section 5.

2   Physical Layer Impact in Communications

The characteristics of satellite RF propagation will determine essential performance parameters for satellite network links: bit error rate, delay or availability, and will affect bit rate adaptation in some satcom systems, such as DVB-S2/RCS. The most representative parameters describing these impacts include: bit error rate (BER), channel capacity and the propagation delay.

2.1   Bit Error Rate and Channel Capacity

In digital satellite communications, the error rate of an incoming data flow is measured by the link BER, which represents the ratio of the number of erroneous received bits over the number of bits transmitted from the communications channel. If the link BER is high, it may exceed the channel coding scheme's capacity of error. For example, in a DVB-S2 system disrupted by heavy rain, the received BBFRAME may have too many erroneous bits, which will be detected by the FEC, and therefore, the frame may be dropped by the receptor. If this situation occurs for successive frames, the channel will be considered blocked or in a state of failure. The calculation of BER in digital communications is directly related to the Carrier power to noise power spectral density ratio (C/N_0) of the link, and the modulation and coding scheme (MODCOD) used for RF communications (where C/N_0 is dependent on the system configuration and the external physical media of transmission, and the MODCOD is only dependent on the system design). The choice of the MODCOD will also determine the effective transmission rate through the communications channel, which is relevant to the whole network performance.

General Link Analysis. C/N_0 shows the ratio of the power of the signal received to the power density of the received noise that interferes in the signal demodulation and is an essential parameter in describing the quality of communications.

To analyse the total C/N_0 (or end-to-end C/N_0 in satellite communications with transparent payloads) [2], there are many parameters to consider, as shown in the following table:

Table 1. Parameters affecting the link budget

Earth station	Satellite	Channel
• Geographical location: Rain fade statistics, Satellite look angle, Satellite EIRP in the direction of the earth station, Earth-Satellite Path loss • Transmit antenna gain G_u + Transmitted EIRP • Receive antenna gain – *related to G/T* • System noise temperature – *related to G/T* • Intermodulation noise - *related to total C/N_0* • Equipment considerations (cross polar discrimination, filters) *related to link margins*	• Orbital position • Tx and Rx antennas - *related to EIRP and G/T* • Power - *related to EIRP* • Transponder gain and noise characteristics - *related to EIRP and G/T* • Intermodulation noise - *related to total C/N_0* • Interference level (C/I) that can be decomposed in inter-system and intra-system interference	• Frequency - related to link margin and path loss • MODCOD - related to required C/N_0 • Intersystem noise / interference

In particular, the main parameter that may dramatically change the C/N_0 throughout time is the attenuation of the channel, which is mostly affected by the weather conditions, i.e. hygrometer absorption due to rain or ice. In mobile systems, shadowing, and multipath propagation also represent important factors, which may eventually decrease the total C/N_0 drastically. Thus, the attenuation of channel due to hygrometers and mobile conditions are key parameters in the representation of a channel model.

Segmented Analysis. In the case of satellite communications, the transmitted signal traverses several segments. It starts from a ground antenna transmission directed to the satellite reception antenna and then crosses the satellite chain, and ends with satellite transmission towards the ground terminal antennas followed by reception in ground (see figure 2). Interference from external noise sources and the intermodulation noise produced in the High Power Amplifier (HPA) in the satellite cannot be neglected for the total C/N_0 calculation.

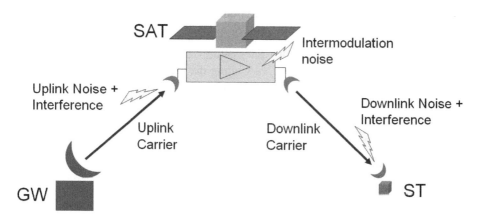

Fig. 2. Segments of satellite communications and noise sources in the global satellite scheme in transparent case

Finally, there is an expression to obtain the total C/N_0 [3], as presented in equation 1, which summarises the end-to-end communications quality:

$$(C/N_0)_T^{-1} = (C/N_0)_U^{-1} + (C/N_0)_D^{-1} + (C/N_0)_{IM}^{-1} + (C/N_0)_I^{-1} . \tag{1}$$

, where $(C/N_0)_U$ is the C/N_0 for the uplink, $(C/N_0)_D$ for the downlink, $(C/N_0)_{IM}$ for the intermodulation, $(C/N_0)_I$ for the interference and $(C/N_0)_T$ for the total Carrier power to noise power spectral density ratio.

Remark: Note that the equation above only applies to transparent satellite-based systems. For the regenerative case, link quality (in terms of BER, for example) is computed independently, per segment. The complete BER consists in, as a first approximation, the addition of BER_U and BER_D, which are independently relative to $(C/N_0)_U$ and $(C/N_0)_D$.

2.2   Delay

In a general data transmission scheme, there exist two sources of delay: transmission delay, which is the time needed in the transmitter to convert a flow of data into a concrete sequence of symbols and push them into the physical medium, and propagation delay, which implies the transportation of those symbols across the physical medium from transmitter to receiver.

In satellite communications, the transmission delay does not present any particular difference compared to regular terrestrial RF systems, however the physical propagation of electromagnetic waves in satellite communications implies a significant delay, due to large distances between the satellite and the earth station antennas. Considering all cases of satellite orbits, a delay model may be seen as a combination of:

- The source ground terminal to source satellite propagation delay (t_{up})
- The Inter-satellite link propagation delays (t_i)
- The destination satellite to destination ground terminal propagation delay (t_{down})

The propagation delays t_{up} and t_{down} are dependent on the distance between the ground antennas and the satellite. In case of GEO satellites, this value will be almost fixed. In case of HEO/LEO, t_{up} and t_{down} will vary along the orbit translation and it could be easily computed as an orbital position problem and distances determination. Also, the calculation of t_i in case of satellite constellations requires modelling of inter-satellite links, which may be more complex [4].

3   Satellite Network Emulator - PLATINE

The aim of this study is oriented to integrate a logical structure for physical layer representation over a satcom system emulator. In our case this platform is PLATINE, but the aim is to propose a general structure that could also be valid for other platforms [5][6]. The working modes considered include two main scenarios: a transparent satellite case, implemented in star topology, and the regenerative case, where a mesh structure is used and only a single hop is needed to interconnect two satellite terminals.

In the architecture of PLATINE, each network element, such as Gateway (GW), Satellite Terminal (ST) or Satellite (SAT), is emulated in a dedicated machine and all of them are interconnected with LAN Ethernet. The satellite core network is emulated by the SAT machine as link emulator and the Network control center (NCC)/GW machine as bandwidth manager (Demand Assignment Multiple Access – DAMA).

Concerning the physical layer representation in PLATINE, two different modules have been integrated in the past. The first implementation included the generation of error bursts (a group of consecutive erroneous bits in the data stream), using statistical parameters or loading pre-calculated error files. In addition, MODCOD and Dynamic Rate Adaptation (DRA) scheme profiles of each terminal could also be loaded from an externally generated simulation. This implementation was used to analyse the impact of adaptive capabilities on DVB-S2/RCS systems when affected by rain cells. The second implementation proposed real-time emulation of channel attenuation to determine the C/N_0 of end-to-end links according to theoretical attenuation models.

This approach lacked the segmented analysis (independent analysis of each link from any ground station to the satellite and the satellite chain link), which is essential for a representative evaluation of satellite network performance. It also lacked interaction with the waveform and channel coding chosen for satellite transmission.

This study re-uses this previous work and the principle of channel emulation, while reformulating the logical structure and settling a more flexible analysis of C/N$_0$ based on link segmentation.

4   Physical Layer Model

The aim of the implementation is to create a common module in all the machines of the emulator to represent the impact of the physical layer on the satellite network. Regarding the architecture of PLATINE divided in blocks, an independent module for Physical Layer functions is introduced between the DVB-RCS block and the Satellite Carrier block. This module is in charge of routing frames through the emulation network. The implementation of Physical emulation in a new block enables easier and cleaner integration, providing independence for external development and remaining open to wider schemes and new implementations of physical layers (figure 3). Internally, the Physical Layer block presents two inner modules referred to as Channels that manage the state of each segment link, e.g. downlink, uplink, or satellite segment. Since PLATINE architecture follows an object-oriented approach, each Channel is represented by a class.

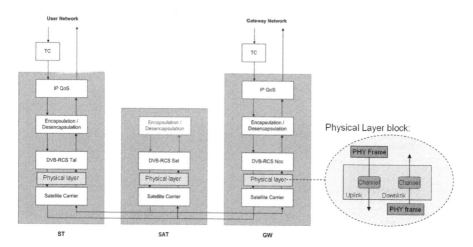

Fig. 3. Block architecture in PLATINE and diagrams of Physical Layer block in different machines

The main function of Channel objects is to manage the flow of incoming or outgoing frames of each machine. Their objective is to reproduce errors and delay when necessary, and provide information about the C/N$_0$ of the links. In fig.4, the Channel structure presents five associated attribute classes; these are virtual classes. Following this approach, a logical structure has been defined to keep coherent Channel classes among the different machines and to allow the implementation of different models for each attribute class, i.e. Attenuation model may be implemented by Rain model or LMS model or, for example, Delay class may be implemented by a LEO or GEO model of delay.

The five attribute classes for Channels are:

- *Attenuation class* represents the attenuation state of one link, and stores the value in the attribute Attenuation (dB). It is further implemented by different channel models, i.e. Rain model, LMS model or other models.
- *Nominal Condition class* represents the best C/N_0 possible in the link. For example, for Uplink and Downlink, that value corresponds to clear sky C/N_0. In the case of satellite segment channels, C/N_0 corresponds to the best $(C/I)_{IM}$ that can be achieved using the amplification and processing chain.
- *Waveform class* manages and obtains information about the current modulation and coding scheme used for each frame. Therefore, other attribute classes, such as Error Insertion, may use this information to complete their tasks.
- *Error Insertion class* (optional) generates and inserts erroneous bits into incoming frames depending on the $(C/N_0)_T$ obtained and the Waveform. In case of transparent payloads, this class is only present in the receiving channels (Download channels) in the ground stations. Otherwise, it is present in the satellite segment in case of regenerative payload. Two models are initially proposed: Single bit error, which calculates a reference BER and inserts erroneous bits accordingly, and Error Gate, which drops every incoming frame if the $(C/N_0)_T$ is insufficient compared to a given threshold value.
- *Delay class* inserts a time elapse between the reception and reemission of the frame. The duration of this delay will depend on the implementation model chosen.

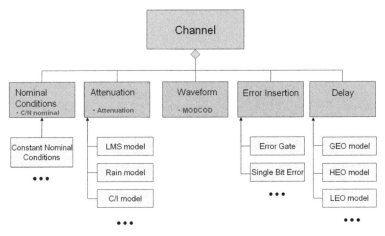

Fig. 4. Class structure for Channel module

4.1   Principles

In each frame transmission from one ground station, such as ST or GW, to another ground station, the "Physical Layer" block will determine how to affect the communications, blocking the frame transmission or inserting erroneous bits if necessary. One main element must be determined for each end-to-end frame transmission: $(C/N_0)_T$, calculated as the combination of all the segments the frame has been through, using equation 1. The C/N_0 of each segment (uplink, downlink or satellite segment) is calculated in each Channel module as:

$$(C/N_0)_X = (C/N_0)_X{}^{Nominal} - Attenuation .$$
(2)

4.2 Execution Steps

In the case of transparent communications, the sequence for the Physical layer function implies three stages, as shown in fig.5. First, the $(C/N_0)_U$ is calculated in the Channel object for uplink in the emitting ground station and this information is inserted in a special header in the frame, obtaining a PHY-frame. The calculation of C/N_0 for the uplink requires the emulation by the Channel object, calculating the attenuation in that given time and subtracting it from its nominal C/N_0 value. Thereafter, the PHY-frame is routed through the SAT machine, where an additional $(C/N_0)_{IM}$ due to intermodulation noise, is optionally added to the PHY-frame. Finally, the frame arrives at the destination machine and $(C/N_0)_D$ is calculated by the downlink Channel object. Then, gathering $(C/N_0)_U$, $(C/N_0)_D$ and $(C/N_0)_{IM}$, the parameter $(C/N_0)_T$ is obtained using equation 1 and it may be used to calculate error effects.

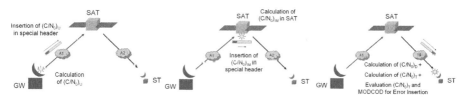

Fig. 5. Distributed calculation of $(C/N_0)_T$ in transparent case

In the case of regenerative communications, the evaluation of $(C/N_0)_T$ is also performed when the frame arrives at the SAT machine and errors effects are also introduced.

4.3 Results

To validate an initial implementation of physical layer over PLATINE, several tests have been developed. Rain and ONOFF models of attenuation over ST links have been applied. Errors have been emulated by frame corruption when the $(C/N_0)_T$ value did not reach a minimal threshold value. For example, Iperf tests have been used to generate UDP traffic at constant bit rate (CBR) and to internally validate the correct calculation of $(C/N_0)_T$ for end-to-end frame transmissions. Results has been compared to an ideal model based on NS-2 simulations. An example of ONOFF channel is shown below, where a loss of bit rate efficiency may be appreciated compared to the ideal model, due to the protocol overhead considered in the Platine emulation. Also, another difference between both platforms is that channel disruption has been implemented as frame loss during physical propagation in the NS-2 simulation whereas frame corruption has been applied at end-user level in our physical layer over Platine.

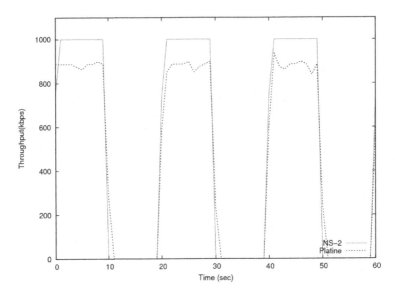

Fig. 6. Performance comparison between Platine and ideal simulation with NS2. Test: UDP traffic received by the GW from ST, with CBR traffic generation of 1 Mbps over a 1Mbps CRA channel in Return link, and affected by periodic disruptions of up/downlink every 10 seconds. DVB-RCS/ATM/AAL5 protocol stack implementation over Platine yields a lower traffic throughput due to the overhead introduced.

5   Conclusion

First, a thorough analysis of RF satellite communications has been schematized and simplified into logical modules adapted to the network emulation case. Here, the insertion of errors based on $(C/N_0)_T$ calculations and delay has been chosen as the most representative impacts of the physical layer on the network emulator.

The definition of a logical and class structure for the physical layer presented some challenges, such as setting up a modular structure capable of providing elementary models of the physical layer, while being sufficiently adequate for future complex models, like new models of attenuation.

The main achievement of the design lies in the establishment of segmented analysis of end-to-end $(C/N_0)_T$, since it distributes the individual link C/N_0 calculation among all the emulator's machines and improves the computation efficiency of the system and avoids overloading certain machines.

At the moment, only basic tests have been performed to demonstrate the correct functioning of the emulator. Errors have been implemented with blocking states, when $(C/N_0)_T$ was below a certain threshold.

The next version of Physical layer implementation will introduce error effects directly translating the $(C/N_0)_T$ into Bit Error Rate (BER), which assures a gradual degradation of communications with the decrease of $(C/N_0)_T$. This feature will ease the QoS evaluation of real-time services such as voice communications over IP or video conferences. Furthermore, the $(C/N_0)_T$ value calculated in end-to-end communications will be used to test the adaptation of upper layers with changes in modulation and coding schemes, to improve robustness against errors and modify the bit rate.

Other future improvements include dynamic changes in execution time, such as the introduction of transition profiles of Channel models, e.g. change from heavy Rain conditions to a clear sky model. Finally, new sources of degradation impacting the total C/N_0 calculation shall be taken into account, such as the intermodulation effect in the satellite segment or the modelling of interference noise.

References

1. ETSI, Digital Video Broadcasting (DVB); Interaction Channel for Satellite Distribution Systems, DVB-RCS standard, EN 301 790
2. Richharia, M.: Satellite communication Systems, Design Principles. McGraw-Hill Inc., US (1995)
3. Maral, G., Bousquet, M., Sun, Z.: Satellite Communications Systems: Systems, Techniques and Technology, 5th edn. John Wiley and Sons (2009)
4. Goyal, R., Kota, S., Jain, R., Fahmy, S., Vandalore, B., Kallaus, J.: Analysis and Simulation of Delay and Buffer Requirements of satellite-ATM Networks for TCP/IP Traffic, OSU Technical Report (1998), http://www1.cse.wustl.edu/~jain/papers/satdelay.htm
5. Baudoin, C., Arnal, F.: Overview of Platine emulation testbed and its utilization to support DVB-RCS/S2 evolutions. In: Advanced Satellite Multimedia Systems Conference (ASMA) and The 11th Signal Processing for Space Communications Workshop (SPSC), 5th edn., September 13-15, pp. 286–293 (2010)
6. Pinas, D.F., Morlet, C.: AINE: An IP network emulator. In: 10th International Workshop on Signal Processing for Space Communications, SPSC 2008, October 6-8, pp. 1–7 (2008)

Evaluating Web Traffic Performance over DVB-RCS2

Raffaello Secchi, Arjuna Sathiaseelan*, and Gorry Fairhurst

Electronics Research Laboratory (ERG),
University of Aberdeen, AB24 5UE, Aberdeen, UK
{raffaello,arjuna,gorry}@erg.abdn.ac.uk

Abstract. The web has undergone a radical change over time. Changes not only in the volume of data transferred, but also the way content is delivered to the user. Current web server architectures are often highly distributed and adapted for user interaction, with transactions characterised by multiple connections to multiple servers. This paper discusses the implication of this new web on next generation two-way satellite systems. It seeks to answer the question of whether classical resource provisioning remains suitable for this traffic. It first presents a more representative simulation model that captures the key features of modern web traffic. It then uses simulation to evaluate the performance over the second generation of DVB-RCS, assessing the impact on performance for a range of bandwidth on demand methods. This paper may be used to formulate recommendations for how to support web traffic in DVB-RCS2.

Keywords: Web Traffic, HTTP modelling, DVB-RCS capacity categories.

1 Introduction

In the last decade the Word Wide Web (WWW) has radically changed both in terms of complexity and usability. In May 2010 Google published statistics [1] about the size and composition of web pages. This showed many current web pages are significantly more complex compared to when version 1.1 of the Hypertext Transfer protocol was specified HTTP/1.1 [3]. This increase in complexity has been accompanied by a substantial increase in HTTP traffic, with a current average web page size at least ten times larger than a decade ago [3]. Since web is a key service carried by satellite, it is important for the satellite community to understand the implications of this new web, especially the implications on emerging next generation systems, such as the new DVB-RCS2 (Digital Video Broadcasting Return Channel via Satellite ver. 2) standard[4]. This requires a

---

* Raffaello Secchi was supported by Astrium UK, subsidiary of the European Aeronautic Defence and Space (EADS) company. Arjuna was supported by the Satellite Internet for Rural Access (SIRA) project funded by the RCUK Digital Economy Programme.

P. Pillai, R. Shorey, and E. Ferro (Eds.): PSATS 2012, LNICST 52, pp. 148–155, 2013.

detailed study of web characteristics, development of a representative simulation model and evaluation of the interactions with the satellite system.

Past work has explored the interactions of TCP and satellite bandwidth allocation methods [2,13,14,15]. These papers have considered server workload models to explore the effects of web over the first generation DVB-RCS system. The interaction of HTTP over DVB-RCS was studied in detail in [15]. Although these papers presented detailed web models, they may not be able to capture the essential nature of the present web. They lack: (i) Support for a distributed server architecture, employed by the majority of popular web services, (ii) Modelling of web browser configurations (e.g. number of simultaneous connections per server and proxy-based approaches), (iii) Support for the network-layer IP signalling required to download a web page, such as the domain name service resolution (DNS). This raises concern about the effectiveness of the models for evaluating HTTP performance and use to explore design decisions for future satellite systems.

This paper is a first step to understand the implications of current web on satellite networking. We define a web model that supports multiple connection per server to fetch the objects that we suggest closely reflects the current parameters of web traffic. Then we use the model to simulate a web transfer across a satellite network with the classical Bandwidth on Demand (BoD) allocation methods. Our results show interesting properties of dynamic allocation to web flows.

The remainder of the paper is organised as follows: Section 2 describes the key characteristics of the web and provides a brief overview of our proposed simulation model. Section 3 discusses the performance of web over DVB-RCS2, followed by conclusions.

1.1   The World Wide Web

HTTP [8] is the de facto protocol for delivery of web pages. In HTTP, a web browser acts as a client, while a server application hosts the web site. The client submits a HTTP request message to the server. The server returns a response message to the client, which contains completion status information and may contain content requested by the client. A web transfer usually consists of a sequence of these request-response transactions, possibly to multiple servers, in order to transfer the set of objects forming a web page.

HTTP was initially standardised as HTTP/1.0 [8] and revised in HTTP/1.1[9]. While HTTP/1.0 used a separate connection to the same server for every request-response transaction, HTTP/1.1 can reuse the underlying transport connection multiple times, (process known as *HTTP pipelining*), to deliver the objects at the same server for a given web page. Since establishment of new TCP connections incur considerable delay, web pages transferred using HTTP/1.1 experience less latency. HTTP/1.1 also introduced the *chunked transfer* method to allow an object to be sent as a continuous stream and the *byte serving* method, allowing a server to transmit any portion of a resource. Currently these methods are used extensively in web transfers.

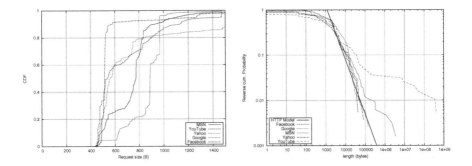

Fig. 1. CDF of HTTP request size from a selected pool of popular websites and reverse CFD of the relative response size

1.2 Characterisation of Web Traffic

Statistics published by Google [1] show that the overall size of a webpage is 320 KB on average. This study found that typical web pages consisted of many objects (around 40) that are downloaded in parallel from 7 connections on average. HTTP requests often includes also the optional body, which carry user information (the Cookie) increasing substantially the size of the request packet. The size of the response i.e. the downloaded object size follows a Pareto distribution with a mean object size of 7187 bytes and a shape parameter of 1.2 [1].

To explore these findings a set of measurements were performed using a pool of popular web sites. Figures 1a and 1b respectively show the cumulative distribution function (CDF) of the size of requests and the reverse CDF of the size of responses of some websites we considered. We observed that an HTTP request is typically 200 bytes longer than a decade ago [5], mostly due to Cookie insertions, and much more variable in length. Requests can be larger than a TCP maximum segment size (MSS). On the other hand, the characteristic distribution of HTTP responses has been preserved and can be well approximated, as predicted in [1], by a Pareto distribution with average 7.1 kB and shape parameter 1.2 (see Figure 1(b)). However, Figure 1(b) shows that the Pareto model may not be suitable when using HTTP streaming. For example, a YouTube session may embed video-clips in HTTP Streaming that would not fit this model.

Modern web scenario is characterised by heterogeneity of web technologies. Different web browsers use different policies when downloading webpage objects. When a webpage is accessed by clicking on the URL, the client sends a HTTP GET request to the server requesting the main file (index.html). Once the main file is downloaded, the client then sends requests to download the objects that are indexed by this file. These requests are sent concurrently, limited by the allowed number of concurrent connections. This number varies between browsers. RFC 2616 [9] originally stated that clients that use persistent connections should limit the number of simultaneous connections that they maintain to a given server. A single-user client should not maintain more than 2 connections with any server or proxy. However, Internet Explorer, IE 8, allows 6 concurrent connections whereas

IE 7 and earlier allowed only 4 connections [10]. Opera developers suggest a maximum of 8 connections per server. Mozilla Firefox allows a maximum of 15 connections per server [11]. Google Chrome uses a maximum of 6 parallel connections per group.

1.3   Characterising the Page Based Simulation Mode

A preliminary survey of simulation models found that many web traffic genera-tors are still based on outdated parameters and have not been updated to reflect the characteristics presented in the previous section. For example, the web mod-els available in the popular ns-2 simulation package [3,5,6,7] may not be able to capture the real web scenario. To mitigate this problem we developed a web traffic model based on [1] and our laboratory analysis. Due to lack of space, we could not report our entire analysis (Figure 1 reports the request/response sizes). However, the following text summarises the steps necessary for a client to retrieve a web page.

A HTTP transaction starts with a DNS request/response. Initially, a HTTP request is sent for the first main object. Once received, a set of concurrent connections are opened (the TCP connection used to transfer the first object may be reused). Subsequently, the client requests a DNS lookup for each server it needs to access. When the connections get the responses for their requests, they may be reused:

1. The client first sends a DNS request to determine the IP address of the first server it must connect. The client receives a DNS response. The RTT is assumed to be twice the simulated one-way delay.
2. Once the DNS response is received, the client sends a HTTP GET request to the first server. It returns a HTTP response (the main object).
3. Once a client receives the main object, it tries to fetch all the related objects from the server (or multiple servers). It sends DNS requests for multiple servers (if any). Our model uses a random number of servers (up to 7). We also model a random number of objects at each server.
4. Once DNS responses have been received, the client opens concurrent con-nections to each server. An inter-server delay represents the time between requests to different servers, modelling client processing.
5. The client closes all TCP connections when all the objects have been down-loaded for a webpage[1].

2   Performance of Web Traffic with Capacity Categories

This section describes a set of simulations to evaluate the interaction between the capacity categories of DVB-RCS2 [4] and the presented web traffic model.

---

[1] If further objects are to be fetched from the first or main server, the client opens parallel TCP connections to fetch these objects. The TCP connection used to fetch the main file is reused.

This uses an ns-2 simulation model of the DVB-RCS2 [16]. The simulator models an RCS terminal (RCST) and the corresponding gateway that provides Internet connectivity to the satellite network. The RCST make a capacity request (CR) for a set of TCP flows from the network control centre (NCC). The NCC makes corresponding allocations, providing return link (RL) capacity towards the gateway.

An HTTP server may be accessed by an RCST via a gateway. In this case, the RL transports mostly HTTP requests and acknowledgements (ACKs). In an alternate use a web server may be provided at an RCST. The former is typical for traditional broadband satellite access using a star network topology, whereas the latter is possible when an RCST is uses as a gateway in a regenerative satellite system or as a mini-gateway in a star system enabled to support Mesh connectivity. This second scenario is used to provide a good illustration of the impact of the BoD dynamics on web service performance.

The web server receives requests for a complex webpage (43 objects) and returns the web contents through the RL. In a typical star network, the NCC incurs an allocation delay, i.e. the delay between a CR and its corresponding assignment, is about two satellite round trips (about 640 ms). The CR request period in our simulations uses one request per second when traffic activity is detected. The satellite frame period was 26.5 ms and consisted of carriers with 16 timeslots. Packets were encapsulated using a return link encapsulation (RLE) [4] with a burst size of 53 bytes. The total transmission rate is 256 kb/s.

The allocation method used a combination of RCS capacity categories. The RCST sends Rate Based Dynamic Capacity (RBDC) requests every second averaging the input traffic over the past interval. A smoothing filtering with parameter $0 \leq \alpha \leq 1$ can be also applied to the rate samples. RBDC(α) denotes in our graphs an RBDC method with filtered samples. The RCST may use a Volume Based Dynamic Capacity (VBDC) request in addition to the RBDC request in the request message. The NCC only allocates the minimum amount of slots to satisfy a request in each transmission burst time plan (TBTP).

The client used a transport based on TCP New-Reno SACK. TCP parameters were default with a packet size of 1500B (including the TCP/IP header). The simulations considered three web-pages whose object sizes were extracted from the previously described empirical distribution. The total web page sizes were 212, 178, and 315 KB.

2.1   Performance Analysis

Figure 2 reports the completion of the web page transfer and the corresponding allocation efficiency. The efficiency is calculated as the amount of capacity used (in bytes) including encapsulation overheads with respect to the amount allocated, and it does not consider the capacity allocated after the completion of a web transfer until the RCST channel becomes idle.

The number and the size of objects forming a web page can vary. Apart from the BoD type and page size, many other factors influence the transfer completion time (TCC), such as the burst size, carrier bitrate, order of scheduling of HTTP

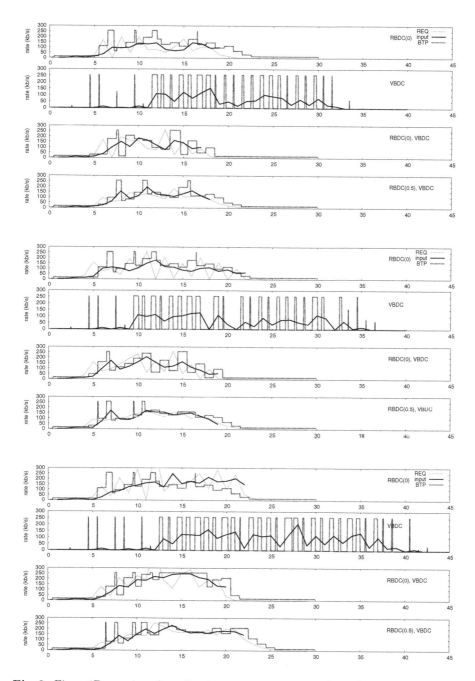

Fig. 2. Figure1Dynamics of application transmission rate (input), capacity requests (REQ), and allocated rate (TBTP) for four allocation methods (RBDC(0), VBDC, RBDC(0) + VBDC, and RBDC(0.5) + VBDC). Three HTTP test sets with different distributions of object size.

Table 1. Completion time and allocation efficiency (bytes used/allocated) for web transfers for different BoD methods

Dataset	RBDC(0)	VBDC	RBDC(0), VBDC	RBDC(0.5), VBDC
Set 2 (178 kB)	18.6 s, 0.86	30.8 s, 0.99	16.7 s, 0.81	17.4 s, 0.77
Set 1 (212 kB)	21.6 s, 0.87	33.8 s, 0.98	18.4 s, 0.85	17.8 s, 0.82
Set 3 (315 kB)	21.7 s, 0.86	39.8 s, 0.99	19.2 s, 0.85	21.5 s, 0.85

requests at the client, the encapsulation efficiency, etc. This makes it difficult to formulate precise recommendations. Despite this, our analysis highlights some important results:

RBDC allows shorter completion times compared to the VBDC. VBDC tends to request and allocate capacity in bursts (the allocation pattern is ON/OFF). This increases the RTT seen by TCP and slows down the transfer. However, the VBDC efficiency is nearly ideal (see Table 1), since VBDC tends to request the exact amount of capacity that needs to be allocated for the queued traffic at an RCST.

A combination of RBDC and VBDC, whether or not the RBDC requests are pre-processed by a filter, provides better performance than RBDC or VBDC alone.Combining RBDC and VBDC leads to a lower utilisation of the requested bandwidth with respect to VBDC. For instance, the median of the efficiency of RBDC was observed to be between 83% and 94%, and between 77% and 85% for the combined method. This is probably due to the high variability of web traffic rate during the allocation period. The compound method produces performance better than RBDC alone, and can be used to trade latency for efficiency.

3   Conclusions

Changes in HTTP usage have been accompanied by an increase of web document complexity and a transition from single server domains to distributed architectures. This calls for a revision of the models used to evaluate performance for web traffic.This paper proposes a model that captures the characteristics of the HTTP request/reply transaction and a multiple-server architecture.

This model was to simulate HTTP performance with standard bandwidth allocation mechanisms over a DVB satellite link. Our results show that the dynamics of the allocation for a web flow can be significantly affected by the allocation method. We found that the latency of the transmission of a web page is less when RBDC and VBDC were combined rather than using them alone. The performance gain is clearly dominated by the allocation pattern, which is much more stable when RBDC and VBDC work together.

Our research also uncovered pathologies in the standard allocation mechanism that must be considered when designing new allocation strategies. Future research will extend this to include a wider range of allocation mechanisms and provide recommendations for efficient transport of web traffic.

References

1. Ramachandran, S.: Lets Make the Web Faster, Google (May 2010),
   http://code.google.com/speed/articles/web-metrics.html
2. Sooriyabandara, M., Fairhurst, G.: Dynamics of TCP over BoD satellite networks.
   Int. J. Satell. Commun. Network. 21, 427–449 (2003)
3. Cao, J., Cleveland, W.S., Gao, Y., Jeffay, K., Smith, F.D., Weigle, M.C.: Stochastic
   Models for Generating Synthetic HTTP Source Traffic. In: IEEE INFOCOM, Hong
   Kong (March 2004)
4. Digital Video Broadcasting (DVB), Second Generation DVB Interactive Satellite
   System (RCS2): Part 1: Overview and System Level Specification, DVB Document
   A155-1 (March 2010)
5. Mah, B.A.: An Empirical Model of HTTP Network Traffic. In: IEEE INFOCOM,
   Japan (April 1997)
6. Floyd, S., Paxson, V.: Difficulties in Simulating the Internet. IEEE Transactions
   on Networking 9(4), 392–403 (2011)
7. Abrahamsson, H., Ahlgren, B.: Using Empirical Distributions to Characterise Web
   Client Traffic and to Generate Synthetic Traces.In: IEEE GLOBECOM, San Fran-
   cisco (December 2000)
8. Berners-Lee, T., Fielding, R., Frystyk, H.: Hypertext Transfer Protocol HTTP/1.0,
   IETF RFC 1945 (May 1996)
9. Fielding, R., Gettys, J., Mogul, J., Frystyk, H., Masinter, L., Leach, P., Berners-
   Lee, T.: Hypertext Transfer Protocol – HTTP/1.1, IETF RFC 2616 (June 1999)
10. AJAX - Connectivity Enhancements in Internet Explorer 8,
    http://msdn.microsoft.com/en-us/library/cc304129(VS.85).aspx
11. Network. http.max-connections, MozillaZine,
    http://kb.mozillazine.org/Network.,http.max-connections-per-server
12. Hernandez-Campos, F., Jeffay, K., Smith, F.D.: Tracking the Evolution of Web
    Traffic: 1995-2003. In: IEEE/ACM MASCOTS 2003, Orlando, Florida, (October
    2003)
13. Roseti, C., Kristiansen, E.: TCP behavior in a DVB-RCS environment. In: 24th
    AIAA International Communications Satellite Systems Conference (ICSSC), San
    Diego (June 2006)
14. Roseti, C., Kristiansen, E.: TCP Noordwijk: TCP- Based Transport Optimized for
    Web Traffic in Satellite Networks. In: 26th International Communications Satellite
    Systems Conference (ICSSC), San Diego (June 2008)
15. Luglio, M., Roseti, C., Zampognaro, F.: Improving performance of TCP-based
    applications over DVB-RCS links. In: IEEE International Conference on Commu-
    nications (ICC), Germany (June 2008)
16. Secchi, R.: DVB-RCS2 for ns2 (2011),
    http://www.abdn.ac.uk/eng863/dvbrcs_ns2.htm

On Cloud-Based Multisource Reliable Multicast Transport in Broadband Multimedia Satellite Networks

Alberto Gotta, Nicola Tonellotto, and Emnet T. Abdo

CNR-ISTI, National Research Council of Italy, via G. Moruzzi 1, 56124 Pisa, Italy
{alberto.gotta,nicola.tonellotto}@isti.cnr.it,
emnettsadiku@gmail.com

Abstract. Multimedia synchronization, Software Over the Air, Personal Information Management on Cloud networks require new reliable protocols, which reduce the traffic load in the core and edge network. This work shows via simulations the performance of an efficient multicast file delivery, which advantage of the distributed file storage in Cloud computing. The performance evaluation focuses on the case of a personal satellite equipment with error prone channels.

Keywords: Satellite, Multicast, Cloud Computing, ASM.

1   Introduction

The *Cloud* is a novel distributed platform that provides an abstraction between the computing resource and its underlying technical architecture (e.g., servers, storage, networks), enabling convenient, on-demand network access to a shared pool of configurable computing resources that can be rapidly provisioned and released with minimal management effort or service provider interaction. The increased demand of network resources can not sustain the exponential growth of data dissemination required by the next generation Web-enabled services, since the current Internet business models are shifting toward pervasive and ubiquitous mobile devices (e.g., GPS navigators, smart-phones, netbooks) and new socio-economic models arise from the Web 2.0 (e.g., social networking, context awareness, All-as-a-Service). Some of the current types of applications viable to be ported on Clouds will place very high demands on the network, often requiring the speedy delivery of the same data to multiple destinations (one-to-many communications). In cases of data dissemination from a single content provider to a group of consumers Reliable Multicast Transport (RMT) protocols can be a key factor fostering optimal resource allocation. If we look at the traffic load of a single user only that performs the synchronization of all his devices (smartphones, smart-TVs, laptops, NASs, set-top-boxes, etc.), the convenience of a RMT is outstanding. In particular, when all the personal devices are behind the home router (cable or satellite), only a single multicast flow crosses even the access network, and the replicas affect only the Local Area Network at home. In fact, since the current typical employment of Cloud services for home customers is

P. Pillai, R. Shorey, and E. Ferro (Eds.): PSATS 2012, LNICST 52, pp. 156–163, 2013.

devoted to personal devices synchronization of personal multimedia file, i.e. movies, music, photos, e-books, documents, notes, contacts, apps, bookmarks, etc., the main advantage of using an RMT in a multicast domain is the significant reduction of bandwidth on the backbones, that is, in a Cloud scenario, on the core and edge networks.

In the field of RMT protocols, the primary design goals are providing efficient, scalable, and robust bulk data transfer across possibly heterogeneous IP networks and topologies to a group of users, and, more specifically, to a set of devices. In particular, referring to reliable transport means either imposing a certain level of Quality of Service or providing the confirmed delivery of data, when applications requires the integrity of received data.

The Negative ACKnowledgment (NACK) Oriented Reliable Multicast (NORM) protocol [1] supports a reliable multicast session participation with a minimal coordination among senders and receivers. NORM allows senders and receivers to dynamically join and leave multicast sessions at will, with a marginal overhead for control information and timing synchronization among participants. To accommodate this capability, NORM message headers contain some common information allowing receivers to easily synchronize to senders throughout the lifetime of a reliable multicast session. NORM can self-adapt to a wide range of dynamic network conditions with little or no pre-configuration, e.g., in case of congestion situations on network bottlenecks due to traffic overload. The protocol is tolerant to inaccurate round-trip time estimations or loss conditions that can occur in mobile and wireless networks; it can correctly and efficiently operate even in situations of heavy packet loss and large queuing or transmission delays.

A NORM session is defined within the context of communicating participants over a network using pre-determined addresses and host port numbers in a *connectionless* fashion (e.g., UDP/IP) The participants communicate by using a common IP multicast group address and port number. NORM senders transmit data content in the form of *objects* to the session destination address and NORM receivers attempt to reliably receive the transmitted *object* using NACKs to repair requests, in a confirmed delivery fashion. Moreover, the sender logically segments a transmitted *object* into Forward Error Correction (FEC) coding block. The parity segments may be transmitted proactively, i.e., appended to the information part of the coding block, reactively, i.e., reacting to a repair request, or both, i.e., devoting part of the parity segments to proactive aim and the remaining to reactive one.

The main drawback of a reliable multicast delivery is that the sending agent adapts the throughput in order to not penalize the most impaired receiver (e.g. a personal mobile satellite equipment in severe fading conditions). The other members of a group see this as performance degradation, since the delivery may last for longer. The main overhead in terms of data delivery delay, in case of disruptive channels, is due to the error recovery phase. This is particularly harmful when long latencies are experienced on the channel, as in the case of GEO satellite links, since the retransmission request requires at least twice the channel latency in time units to be satisfied.

Even if no standard protocols have still been defined for RMT, NORM is the best candidate and is going to be finalized as reference standard by IETF-RMT-WG [2].

This work presents the performance evaluation of an Any Source Multicast (ASM) multi-source multicast session, which relies on the architecture of Cloud/Grid distributed parallel storage servers, and which aims at reducing the data delivery delay with a specific advantage for those Cloud applications that would require to achieve a quasi-real-time playing of multimedia files.

2   State of the Art on File Delivery in Grid/Cloud Networks

In the case of unicast communications in Grid networking, GRIDFTP is an extension of the standard File Transfer Protocol (FTP) for use with Grid computing [3]. It is defined as part of the Globus toolkit, by the GRIDFTP working group [4]; GRIDFTP can be used independently, to provide high-speed file transfers. An FTP data transfer is limited by the maximum size of TCP/IP packet and the acknowledged (ACK) reception of each data packet. In WANs, or in Satellite Networks, a simple file transfer based on FTP is affected by latency, as the transmission delay of data packet and the acknowledgements reduces the data transfer rate. GRIDFTP supports parallel data transfer through FTP command extensions and data channel extensions. The use of multiple streams in parallel (see Figure 1) (even between the same source and destination) improves the aggregate bandwidth over a single stream, according to [5]. Data may be striped or interleaved across multiple servers, as in the network disk cache of a distributed parallel storage server or a striped file system.

GRIDFTP includes extensions that initiate striped transfers, which use multiple TCP streams to transfer data that is partitioned among multiple servers. Striped transfers provide further bandwidth improvements over those achieved with parallel transfers.

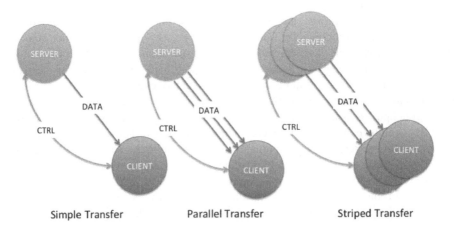

Simple Transfer   Parallel Transfer   Striped Transfer

Fig. 1. Grid FTP transfers

The main concepts that are behind parallel and striped distributions can be summarized as follows:

1. In the case of a single server shared among a large number of receivers, the bottleneck rate may undergo saturation. This leads to a scalability problem that can be only solved by redunding the number of servers.
2. When multiple servers offer different performance levels in term of reliability, due to the network they are respectively behind, two receivers, belonging to the same class of service, may experience different delivery times according to the selected server. This is unfair, in terms of service level agreement, with the respect to customers that have subscribed for the same level of service.
3. With striped transfers from multiple servers, two receivers belonging to the same class should experience at least the same performance, apart from the performance of the different access networks, which are not in charge of the Cloud service provider, but rather of the Internet provider.

According to this backing provided by Cloud/Grid networking, a preliminary investigation on reliable Any Source Multicast (ASM) transmissions is provided in the follow in terms of performance evaluation of NORM protocol over large delay error prone satellite channels.

3   The Case Study

NORM is a protocol centered on the use of selective NACKs to request repairs of missing data. NORM provides for the use of packet-level forward error correction (FEC) techniques for efficient multicast repair and optional proactive transmission robustness [6]. FEC-based repair can be used to greatly reduce the quantity of reliable multicast repair requests and repair transmissions in a NACK-oriented protocol. The principal factor in NORM scalability is the volume of feedback traffic generated by the receiver set to facilitate reliability and congestion control. NORM uses probabilistic suppression of redundant feedback based on exponentially distributed random back-off (BO) timers. This allows NORM to scale well while maintaining reliable data delivery transport with low latency relative to the network topology over which it is operating. The NACKing procedure begins with a random BO timeout in order to avoid the possibility of NACK implosion in the case of sender or network failure. At the end of the BO time, the receiver will suppress its NACK message whether at least one of the following conditions is verified:

1. a NACK message, received from another receiver, equals or supersedes the receiver's repair needs;
2. the receiver detects the sender is sending ordinally earlier blocks (in response to earlier NACKs) than what it is currently pending repair.

Finally, the receiver enters a rest phase, according to another exponentially distributed random timer called hold-off (HO) timer, before starting a new recovery phase, when a new back-off timer will be sorted.

Both BO and HO are function of the round trip time (RTT) of the group (GRTT) of receivers, calculated by the sender/s, which collects the individual RTT of each receiver and assumes the maximum one as that of the group.

When the second condition is verified, the receiver will NACK again in a following repair cycle, after the senders ordinal transmission point will have exceeded the receivers pending repair needs. However, in case of large channel latency as in wireless and satellite technologies, this policy - that avoids multiple repair requests in contiguous repair cycles - stretches the time required to deliver a block of data to the receiving group. Figure 2 highlights through a dashed line the case of a NACK suppression due to the first repair request still pending of resolution.

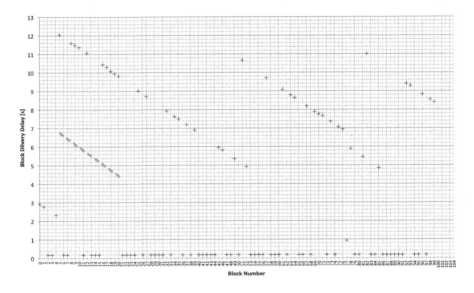

Fig. 2. Data block delivery delay

The simulation environment poses on NS2 v2.34 network simulator patched with the nrl-NORM code [7], which implement the reliable multicast transport according to [1]. The scenario is made up of a certain number of receivers and a variable number of repositories where files are stored. The test in Figure 2 is obtained with only one sender and two receivers, with a GEO satellite channel latency of 125 ms, an average loss probability of 5%, a transmission rate of 256 kbps. A data block, i.e., the FEC code information part, is made up of five segments of 1024 bytes each and one parity segment is devoted to the reactive redundancy.

In Figure 3, 10 consecutive files of 512000 bytes size have been delivered according to the single delivery in Figure 2, in order to depict individual file (object) delivery over time. Each rectangle represents the start and completion of a delivered file. The rectangles overlap because the file delivery of multiple files overlaps each other, i.e., as new data is sent while repairs of the previous file are still occurring.

Looking at the "per file" goodput, it is about 160-180 kbps but the overlapping delivery of the series of multiple file objects comes an average goodput of about 220 kbps, which is not bad for 256 kbps sender rate and 5% packet loss. In fact, each file delivery takes about 24-25 seconds to execute, in each run.

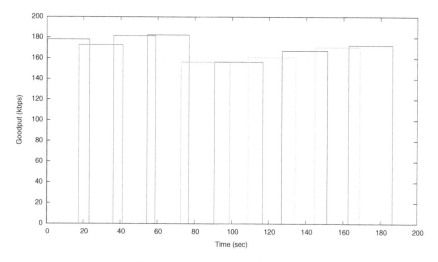

Fig. 3. Object delivery performance

In case of best effort file transfer, the sequential delivery of blocks is not a requirement, and NORM performs quite well. However, in case of on-demand multimedia playing from the Cloud, an out of order delivery of subsequent data block will cause the freezing of the playing. In Figure 2, e.g., block #5 is delayed of about 12 s; this means that after playing 20 KB of the file, the player freezes for more than 12 s, in addition to the 3 s required to sequentially play the earlier four blocks of the sequence.

The only first suppression causes a shifting in time of the following recovery cycles for the pending blocks. This effect is degenerative and incremental with the file size.

Proactive FEC might be a solution, at expense of an overhead of bandwidth that reduces the overall goodput. However NORM does not foresee a control mechanism for proactive FEC adaptation, and, even in case of that, the advantage would stand if the delivery takes a much longer time than that required to the algorithm to adapt to the channel impairments.

A different solution relies on the chance of segmenting the original file into chunks and deliver the object from multiple servers, each of them accounts for delivering a single chunk or few of them. This policy was formerly foreseen by the multicast Internet protocol, which accounts for the ASM as in the case of the Internet Group Management Protocol (IGMP) [8] for IPv4 networks, i.e., one or more sources and multiple receivers; in addition, it stems at the basis of Grid file transfer [3] policies for large file delivery, and has been being inherited by Cloud computing. In fact, over the past few years, Cloud computing has emerged as a new paradigm in advanced computing as a flexible, on demand infrastructure aiming at transparently sharing data, calculations, and services among users of a massive Grid [9].

A file delivery is performed by choosing the number of sources (repositories) according to the number of chunks, in which the file is split. Between the core network, where there are the repositories, and the receivers a satellite bottleneck link play the role of the access network. The satellite link bit rate is set to 1 MB/s and the latency is 125 ms.

Table 1. Performance Evaluation of ASM delivery to 2, 5, and 10 Receivers

Number of Receivers	Chunk Size [KB]	Allocated Servers	Average FDD [s]	Average BDD [s]
2	100	32	7.13	4.30
	800	4	11.97	3.75
	1600	2	14.67	4.44
	3200	1	22.40	7.58
5	100	32	7.68	4.41
	800	4	12.19	4.28
	1600	2	15.23	5.54
	3200	1	24.49	9.43
10	100	32	8.32	4.48
	800	4	12.77	4.37
	1600	2	15.47	5.74
	3200	1	24.75	9.75

In Table 1 we investigate a 3.2 MB file delivery delay (FDD) – a typical MP3 song – segmented into a certain number of chunks, given the number of receivers. The NORM sending agent is configured, in order to organize the chunks into blocks of 10 data segments and 5 additional parity segments are generated for reactive recovery, i.e., no proactive parities are sent, appended to information. In addition, block delivery delay (BDD) is provided in Table 1. The simulations are performed with 2, 5, and 10 receivers respectively. We assume that information segments are released to the application layer by the transport agent through a de-jitter buffer, when all the segments of a received block are successfully retrieved, either when the decoding is required or not. For this reason BDD plays the main role in place of the segment delivery delay.

The simulations reveal that there is an optimal choice of the chunk size, and hence, of the number of allocated servers, independently from the number of receivers. This size is shown in Table 1 approximately between 100 and 800 KB. The reason has to be found in relation to the GRTT, which determines the timings for the BO and HO timers. From the log files, we have experienced that, when the chunk is too short (e.g. 100KB as in Table 1), the recovery phase after a suppression happens when the file delivery is already finished and only the missed packets are waited from recovery phases, before the relative blocks of data are passed to the application layer. By choosing an opportune chunk size (e.g. 800 KB) the number of server is optimized and the BDD can gain even more than 70%, by reducing the average delivery delay per block.

We remark that keeping the BDD as low as possible is one of the main goals in case of on-demand multimedia application, in order to match the user agreement and avoiding the suspension of a live application.

4   Conclusions and Future Works

These preliminary simulations show the benefit of ASM delivery in the context of Cloud networking, when the access network presents long latency. This technique allows reducing both the network load and the delivery delay, adopting a reliable multicast protocol. According to the parallel striped transfer, provided by Cloud computing, the future activities will deepen this investigation and will account for other challenging scenarios with particular interest to the enhancement that network coding could provide in the case of file delivery and broadband multimedia services.

Acknowledgments. The authors thank B. Adamson for his valuable support provided here and in our ongoing research activity on reliable multicast.

References

[1] Adamson, B., Bormann, C., Handley, M., Macker, J.: Negative-acknowledgment (NACK) Oriented Reliable Multicast (NORM), RFC 5740 (November 2009), http.//tools.ietf.org/html/rfc5740
[2] http://datatracker.ietf.org/wg/rmt/charter/
[3] Taylor, I.J.: From P2P to Web Services and Grids - Peers in a Client/Server World. Springer (2005)
[4] http://forge.gridforum.org/projects/gridftp-wg/
[5] Butler, D.: GRIDFT Server Simple Performance Measurements, BBC R&D White Paper WHP178, http://www.bbc.co.uk/rd/pubs/whp/whp178.shtml
[6] [RFC3453] The Use of Forward Error Correction (FEC) in Reliable Multicast
[7] http://cs.itd.nrl.navy.mil/work/norm
[8] [RFC4604] IGMP protocol Using Internet Group Management Protocol Version 3 (IGMPv3) and Multicast Listener Discovery Protocol Version 2 (MLDv2) for Source-Specific Multicast
[9] Tuan-Viet DINH. Cloud Data Management, Bibliography Report, ENS de Cachan (February 2010), ftp://ftp.irisa.fr/local/caps/DEPOTS/BIBLIO2010/Dinh_Viet-Tuan.pdf

Which Transport Protocol for Hybrid Terrestrial and Satellite Systems?

Ihsane Tou[1,2], Pascal Berthou[1,2], Thierry Gayraud[1,2], Fabrice Planchou[3], Valentin Kretzschmar[3], Emmanuel Dubois[4], and Patrick Gélard[4]

[1] LAAS-CNRS, 7, Avenue du Colonel Roche, 31077 Toulouse, France
[2] Université de Toulouse; UPS, INSA, INP, ISAE; LAAS; 31077 Toulouse, France
[3] EADS Astrium, 31 rue des Cosmonautes, Z.I. du Palays, 31402 Toulouse Cedex 4, France
[4] CNES, Centre National d'Etudes Spatiales, 18 Av. Edouard Belin, 31400 Toulouse, France
`{firstname.lastname}@{laas,astrium,cnes}.fr`

Abstract. Satellite systems complement terrestrial networks where the network could not be deployed for technical or economical reasons. Moreover, the natural broadcasting capacity of satellite networks makes it a good companion to terrestrial networks. In this context, future services will be deployed over networks that combine terrestrial and satellite systems. The infrastructure heterogeneity could be problematic, especially because of the delays variation. This article presents the problem from the point of view of the transport layer, the layer directly connected to the application, and compares several solutions to help future service developers using such network configuration.

Keywords: TCP, Hybrid networks, satellite, PEP.

1 Introduction

During the last decade, a wide range of network access technologies have been developed to extend the access to the Internet services. In parallel, the cellular networks, originally designed for voice/telephony mobile services have evolved to offer more services, such as Internet access. The convergence of fixed and mobile services has been achieved and standardized. Also, there has been significant progress on the level of the terminal handset (mobile phone, smartphone and laptop) whose size has been significantly reduced, while providing more capabilities and wireless interface support.

With NGN and 4G architectures, services or applications are designed independently of the underlying access network (wireless, cellular, wired, optical, etc) based on the IP core technology, which is the convergent corner-stone of telephony and data services. The always-on paradigm is conceived as a generalized mobility for user services, allowing seamless service switching across any compliant network access technology.

This means that applications (thus the underlying transport protocols) have to be persistent to the network switching. This is a challenging objective since the access media are heterogeneous and potentially operated by various actors.

P. Pillai, R. Shorey, and E. Ferro (Eds.): PSATS 2012, LNICST 52, pp. 164–173, 2013.
© Institute for Computer Sciences, Social Informatics and Telecommunications Engineering 2013

This paper exposes the problems met by the transport layer and especially TCP with such heterogeneity and compares several solutions to help future service developers using hybrid networks. First, scenarios envisaged for the integration of satellite access systems in future networks, what we call hybrid satellite and terrestrial networks are detailed. Secondly, the performances of the TCP protocol are compared according to different TCP versions, in the specific case of hybrid handover.

2   Hybrid Terrestrial and Satellite Systems

The satellite integration with terrestrial network can be achieved in several manners. Three complementary visions are presented. A **tightly coupled architecture**, an integrated approach in which a given mobile system (3G, LTE, WIMAX) is extended to support the satellite media, as an alternate radio access channel for the mobile user, in a completely transparent way. A "**relay integration**", in which the satellite is integrated within the mobile network infrastructure, not directly at the mobile air interface, but beyond a specific satellite relay enabling the access to the core mobile infrastructure. And a "**loosely coupled integration**", in which a satellite specific interface is added on the mobile terminal enabling the connection to IP data network, through access network. This involves multi-technology (multimodal) mobile terminals, handling several interfaces and running the convenient protocol (specific media access: DVB-RCS+M for instance).

The first two architectures hide the network heterogeneity with a dedicated layer 2 transparent for the upper layers. In the tightly coupled vision, the terminal can be aware of the satellite access; in the relay architecture the heterogeneity is embedded in the network. The loosely coupled architecture needs a third protocol to assure the network change, mobile IP would be the solution, however the network characteristics will be hidden too. Also the heterogeneity is a problem for the transport layer, it appears that an agnostic transport protocol (i.e. it does no know the underlying networks) should be a better solution as the knowledge of the underlying network is a risky hypothesis.

2.1   Impact on the Transport Layer

The hybridization of the networks may cause several troubles with the transport protocols. The firsts occurs when changing network (i.e. leaving a terrestrial network for a satellite network), the others when the transport protocol uses old parameters (Congestion WiNDow size –cwnd-, timers) on a new network (i.e. a TCP connection with its parameters set according to a terrestrial network uses a satellite network with a limited bandwidth and a long delay).

The Handover Model: According to the network specifications and the reception conditions, the handover between two networks can generate service interruption or not. It can vary from several packet losses (soft handover), to short interruption (several seconds), to a long break with network addressing change (mobile IP should resolve this problem). The behavior of the transport layer will depend on this. In this paper, we consider seamless handover and short interruptions that could be implemented using tighten or relay architecture. We exclude mobile IP scenarios as we previously show that such handover in a satellite system is too long for the transport layer (TCP connection reset) [1].

We call the firsts scenarios "make before break" and the last one "break before make" as the adaptation is made after the handover. In the first cases, the new layer 2 is set up before the old one is deactivated. Two links are available during a short period of time when the handover arises. This can provoke a disordered reception of packets and then trigger TCP congestion control algorithms. On the opposite, in the last case, the new link is established only when the old is down. The effects are a risk of disconnection if the disruption is too long (application or TCP timeout), the losses could be interpreted as a congestion and trigger the TCP congestion control algorithms, or an expiration of the TCP retransmission timeout (RTO) that will starve the connection throughput.

Impact of the Parameters Variation: It is obvious that delay, bandwidth and in smaller proportion, the loss rate will vary according to the connected network. The transport layer "guesses" these parameters through *timers* for the delay, *congestion windows* for the bandwidth. Loss rate is slightly different, as it will impact the reaction of the protocol. When changing network, the transport protocol could be duped by the old parameters and do not adapt properly to the new network. Considering TCP, the useless retransmissions could increase because of a too small value of the RTO, or spurious RTO may occur as side effect. **It could increase the unused bandwidth** (i.e. **waste of resources**): TCP may not be able to use the increased bandwidth of new network or, in the opposite, provoke a real congestion if the congestion window is too big for the network capacity.

In this paper we focus on the hybridization problematic. In the following part we briefly summarize the classical TCP versions to compare it when tested in a hybrid context.

3   Transport Protocol Alternatives

TCP [2] is a byte-stream connection oriented protocol. The connection management is handled by the endpoints. TCP uses a sliding window based congestion control scheme: the size of the transmit window is the lesser of the receiver's advertised window size *awnd* and the sender's congestion window size *cwnd*.

The size of *cwnd* is determined by well-known algorithms designed to limit the effects of congestion in the Internet, including slow start, congestion avoidance, fast retransmit, and fast recovery.

The main element that differentiates the TCP implementations is the sending window managment. Many different approaches have been proposed, and published at IETF, without reaching the standard status. Therefore, each operating system may use its own. This section presents a brief overview of different versions that can be interesting because they are well spread and therefore efficient or because they are adapted to a specific domain of application like satellites, mobility or high-bandwidth media.

Reno/New Reno: TCP Reno uses a loss-based congestion control window. It uses the TCP mechanism of increasing the cwnd by one segment per RTT for each received ACK and halving the cwnd for each loss event per RTT. It uses all well-known congestion control [2]. The TCP New Reno improves retransmission during the fast recovery phase of TCP Reno.

Compound: Compound TCP (CTCP) is designed to fit quickly to the bandwidth available while staying TCP friendly. Its particularity is to be both delay-based and loss-based. It manages two cwnds; the classic one like in TCP Reno and the delay based one (*dwnd*) which is used only during congestion avoidance phase.
CTCP is currently implemented over Windows Vista, Seven and Server 2008, Windows Server 2003 and Windows XP 64 bits.

Cubic: The window is a cubic function of time since the last congestion event, with the inflection point set to the window prior to the event.
Cubic does not rely on the receipt of ACKs to increase the window size. Cubic window size is dependent only on the last congestion event.
Cubic TCP is implemented and used by default in Linux kernels 2.6.19 and above.

Hybla: TCP Hybla is a TCP enhancement conceived with the primary aim of counteracting the performance deterioration caused by the long RTTs typical of satellite connections. It adopts the Hoe's channel bandwidth estimate, timestamps and packet spacing techniques.

Westwood: It introduces a modification of the Reno Fast Recovery algorithm. Instead of halving the cwnd after three duplicate ACKs, and fixing the slow start threshold (ssthresh) at this value, TCP Westwood sets the ssthresh at an estimate of the available bandwidth.

Fortunately a number of TCP extensions have been proposed at IETF to improve TCP behavior. Some are general (as TCP options), while others are designed to change a particular TCP behavior or tends to adapt TCP functioning in the conditions of a given media technology (e.g. TCP Sack, Quick-Start), and some are completely specific to the vertical handover purpose (Freeze TCP [3], Fast Adaptation Mechanism [4]).

In the following section, we compare the TCP performances of different real TCP stacks during a handover in a hybrid network.

4   Experimental Test Bed Description

The testbed relies on a satellite emulator (SATEM) developed by ASTRIUM that emulates the handover between satellite and wired network by selecting the available bandwidth, the propagation delay and the Packet Loss Ratio (PLR); this emulator is based on the layer 3 linux network emulator (netem). Client and Server (iperf), a network sniffer (Wireshark) and a TCP proxy (PEPsal) allow to generate, capture and produce the results. The Figure 1 shows a detailed layout of our testbed.

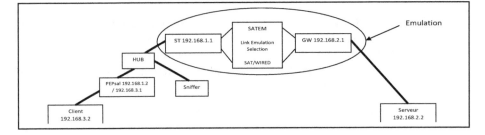

Fig. 1. Testbed description

A Performance Enhancing Proxy has been added to be compared to end-to-end solutions. PEPsal [5] is an integrated PEP scheme based on the TCP-splitting technique that adopts TCP Hybla.

The following metrics has been chosen to evaluate the TCP performances: The *sequence number* evolution of the transmitted packets is useful to monitor the connection efficiency and regularity. It also tells information about packet loss, connection cuts and the amount of data transferred. The comparison of the sequence number evolution is interesting because it can be observed independently of the TCP version used. The *Congestion Window size* (cwnd), the *application end-to-end delay* and *throughput* where also used but not presented in this paper. Theses metrics have been captured and processed using the WEB100 Project and TCPTRACE [6].

The satellite and the wired network parameters on emulated SATEM are:

- Emulated satellite link: Bandwidth 512 Kbps, RTT 500 ± 10ms, P.LR. 10-e5
- Emulated wired link: bandwidth 2 Mbps, RTT 50 ms and free P.L.R.
- The default values of TCP parameters on Linux are used.

We choose to present first results according the variation of only one parameter at one time, to focus on the problems it raise.

TCP Versions over Satellite Comparison

We first choose to compare the TCP versions over the satellite part, even if it has been extensively studied [7], as the performances of Windows Seven Compound version are surprising.

The sequence number evolution graph (a) shows similar performances between hybla, cubic and compound, however compound is more aggressive and benefits rapidly of the available bandwidth without undergoing losses as hybla or cubic. The throughput evolution (b) confirms a better behavior of compound and shows that after 1 minute, the tail of the hybla and cubic curve is the result of large buffers flushing and retransmission management that is not a positive aspect for the application.

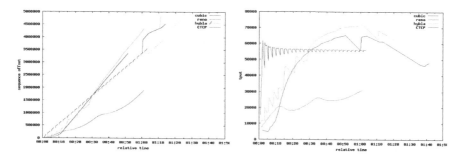

Fig. 2. TCP versions comparison (a) Sequence number (b) Throughput

TCP versions Affording Soft Handover

Our experiences focus on changing one parameter during a handover as described above to see the influence of each parameter during the mobility processes.

Delay variation

The figure 3 (a) shows the evolution of sequence numbers with two handovers. The bandwidth has been set to 2 Mbps during the entire test (60s) and the propagation delay is changing (RTT ~ 500 ms in satellite to 50 ms the wired network). Three phases have been defined, between 0-20 seconds, TCP streams through the satellite network, between 20-40 seconds, they cross the wired network and finally between 40-60 seconds they go back to the satellite network.

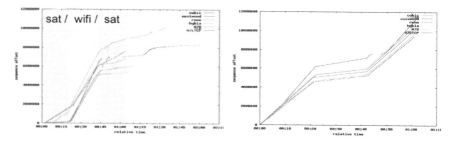

Fig. 3. TCP sequence number evolution with delay (a) and bandwidth (b) variation

In figure 3 (a), as shown previously, hybla, cubic, compound perform better on the satellite part than reference TCP versions (westwood, reno). During the WiFi phase, no significant difference can be observed. The main difference resides when coming back to the satellite network. Cubic affords a bad congestion windows setting and generates losses, hybla too but in smaller proportions. Compound restarts faster and rapidly catches up its delay. The compound dual windows management proves to be effective after a handover.

Bandwidth variation

The figure 3(b) shows the evolution of sequence numbers with two handovers. We set the RTT to 50 ms in the entire test (60s) and vary the available bandwidth (512 Kbps in satellite to 2 Mbps the wired network). The flows undergo three phases: first, they cross the wired network, then they cross the satellite network and finally, they go back to the wired network.

All streams have a quite similar behavior, expect Cubic which suffers at the end of the phase 2. The buffers are over and the cut at the second 43 is a consequence of the aggressiveness of the Cubic mechanism with low bandwidth. Cubic tries to grow up to a maximum bandwidth.

Bandwidth Delay product variation

In order to confirm our results, the same tests have been done using both parameters (bandwidth and delay). Tests have been passed twelve times, and figure 4 shows that compound TCP has a predictable behavior when conditions are stable. The same experience with cubic or hybla does not provide such stability, even if the variation is low.

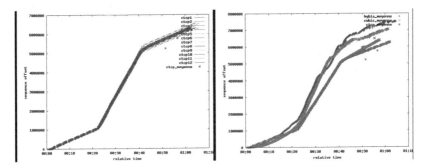

Fig. 4. Sequence number evolution for a dozen (a) CTCP streams and (b) comparison of averages

PLR variation

The variation of the Packet Loss Ratio did not bring significant results, as the reaction of TCP versions should be comparable when affording a loss (considered as a congestion). Cubic obtained the worsts results because of its aggressiveness.

PEP or Not

One of the main question when dealing with specific medium (i.e. satellite) is does we need a PEP or not. Considering our previous result with the new TCP versions over satellite systems, we try to show in this part that splitting or spoofing TCP connection is less mandatory in such configuration. PEP acceleration could still be used for specific applications as big files downloading, but could be omitted for usual Internet browsing. Of course, other PEPs application services, as HTTP caching, are always interresting to reduce the bandwidth usage of the satellite link.

PEP with satellite system

The figure 5 presents the evolution of sequence number of 6 TCP versions through the satellite network. It is clear that the PEPsal solutions are better than end-to-end TCP (in brown is Reno). We notice the exemplary behavior of CTCP (without PEP), which remains stable, little bit under but holds the comparison with the PEP solutions.

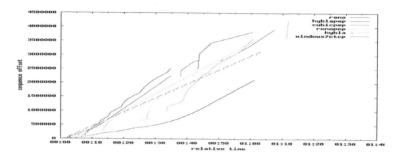

Fig. 5. Sequence number evolution with PEPsal influence over a satellite

PEP with mobility

The figure 6 shows the evolution of sequence numbers with two handovers. In the graph (a), we fix the bandwidth to 2 Mbps during the entire test (60s) and vary the propagation delay (RTT ~ 500 ms and 50 ms). In the second graph (b), we put the RTT to 50 ms during the entire test (60s) and vary the available bandwidth (512 Kbps in satellite to 2 Mbps in the wired network).

The three handovers are defined as previously, from satellite to wired network to satellite.

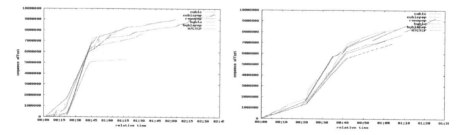

Fig. 6. Comparison between PEPsal TCP and e2e TCP during a handover (delay (a) and bandwidth (b) variation)

The graph (a) clearly shows the superiority of CTCP in Phase 1 and 3. PEPsal has a good result in both phases but with a CTCP advantage. During the transition to the wired network, all flows catch up from the bad performance (especially Cubic), because of the short propagation delay.

In the graph (b), the performance is quite identical between the solutions. A small advantage was noticed during the transition to terrestrial network for the PEPsal.

The Impact of "Break Before Make" Handover

This paragraph compares the behavior of TCP versions when the handover is not error free (break before make scenario). The break was set to 500ms and 1000 ms.

Break (500/1000 ms) before make, fixed bandwidth (2Mbps) and delay variation (500-50-500 ms RTT)

The figure 7 (a) shows the TCP connection behavior when the break takes 500ms. When the streams through the satellite link in the first phase, CTCP confirms the results we had before. The same applies to PEPsal and Hybla, which are just behind. After the first break, at second 20, the restart of all streams is at least about 5 seconds for Cubic, and Hybla PEPsal. CTCP reacts faster and recovers the transmission in about 2 seconds. We get the same conclusion at the second 40, when the second handover occurs.

In the graph (b), when the break takes 1 second, we observed the same behavior after the handover, but with more difficulties for Cubic.

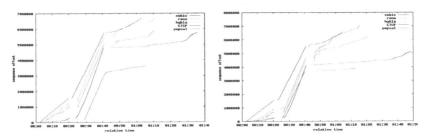

Fig. 7. Sequence number evolution with (a) 500 ms (b) 1000 ms break

Break (500/1000 ms) before make, fixed delay (50 ms) and bandwidth variation (512K-2M-512K bps)

In the figure 8 (a), we see at the beginning of the transmission that Cubic is better than other versions, the delay is low (but with a small bandwidth). During the two handovers, Cubic has a high cut, and restart is difficult. Hybla and PEPsal have naturally the same reaction during handovers. CTCP has a better management of the break, and recovers faster the connection, which takes about 2 seconds.

When the break stays one second, all flows except CTCP, were completely broken after handover (b).

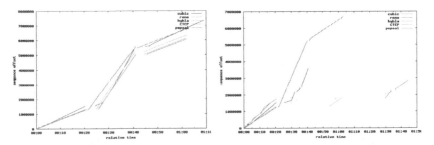

Fig. 8. Sequence number evolution with (a) 500 ms (b) 1000 ms break

5   Conclusions

This paper proposes the evaluation of different TCP versions, using real stacks, in the specific problematic of the handover between terrestrial and satellite networks. The good behavior of the transport layer is crucial to offer quality services over hybrid networks. The surprising conclusion of this study is that windows seven compound stack gives good results, in some cases better than those with an integrated PEP solution PEPsal/hybla, especially over large propagation delays. Tests with the "break before make" showed the difficulties of some TCP restart compared to others, and attest CTCP as the best version for handovers with or without breaks.

References

[1] Arnal, F., Gayraud, T., Baudoin, C., Jacquemin, B.: IP mobility and its impacts on satellite networking. In: ASMS (2008)
[2] Paxson, V., Blanton, E., Allman, M.: TCP Congestion Control, RFC 5681 (2009)
[3] Goff, T., Moronski, J., Phatak, D.S., Gupta, V.: Freeze-TCP: A true end-to-end TCP enhancement mechanism for mobile environements. In: INFOCOM (2000)
[4] Li, D., Sleurs, K., Vani Lil, E., Van de Capelle, A.: A fast adaptation mechanism for TCP veritcal handover. In: ATC (2008)
[5] Caini, C., et al.: Analysis of TCP live experiments on a real GEO satellite testbed. In: Performance Evaluation, vol. 66 (2009)
[6] Shawn Ostermann. TCPTRACE, http://www.tcptrace.org/
[7] Cavendish, D., Tsuru, M., Oie, Y., Kumazoe, M.G.K.: Transmission Control Protocols Evaluation over Satellite Networks. In: International Conference on Intelligent Networking and Collaborative Systems (2010)

A Modulator Interface Protocol
for GSE over DVB-SH

Bernhard Collini-Nocker, Michael Noisternig, and Thomas Soboll

Department of Computer Sciences, University of Salzburg, Austria
{bnocker,mnoist,tsoboll}@cosy.sbg.ac.at

Abstract. The evolution of DVB networks towards IP systems has led
to the introduction of a Generic Stream format in second-generation systems. A Generic Stream Encapsulation (GSE) protocol was designed as
an efficient and extensible IP encapsulation protocol for the DVB-S2
system, and there is growing interest to adopt GSE for other second-
generation DVB carriers. The DVB-SH mobile standard includes provisions for GSE, but a number of adaptations are required for full support.
This paper focuses on the seamless transport of GSE packets over the
DVB-SH distribution network towards modulators. To this end architectural aspects are discussed and a specific protocol is proposed.

Keywords: Modulator Interface Protocol, IP, GSE, DVB-SH.

1 Introduction

DVB-SH (Digital Video Broadcasting - Satellite to Handheld) [2] is a second-
generation mobile broadcast standard designed for universal coverage of multimedia services through a hybrid satellite-terrestrial infrastructure. It inherits
the MPEG-2 Transport Stream (TS) as a means for data transport from first-
generation systems, though it includes provisions for the newer Generic Stream
Encapsulation (GSE) protocol. GSE [1] was originally designed as an efficient
and extensible encapsulation protocol for IP and other network layer protocol
data over the DVB-S2 forward link, but is now increasingly being considered for
other second-generation DVB networks. This reflects an ongoing process of accepting IP as a common transmission technology, thereby extending connectivity
beyond the DVB network and enabling the use of existing IP-based technology.
In DVB-SH the full support for GSE requires a number of adaptations. [5] describes physical and link layer modifications. This paper focuses on the seamless
transport of GSE packets over the DVB-SH distribution network towards modulators. GSE is dependent on the underlying base-band frame format, which is not
uniquely defined among DVB standards and may vary depending on the coding
used. This means that a DVB-based distribution network may not be used "as
is" for the delivery of a transmission multiplex towards broadcast modulators,
and a transparent means of forwarding GSE packets formatted for the DVB-SH
carrier is needed.

P. Pillai, R. Shorey, and E. Ferro (Eds.): PSATS 2012, LNICST 52, pp. 174–179, 2013.

2   Outline of the DVB-SH System Architecture

The DVB-SH system aims to provide universal coverage by combining a Satellite Component (SC), which provides global outdoor coverage, with a Complementary Ground Component (CGC) for cellular-type service in environments where reception via satellite may be impaired. The CGC consists of terrestrial broadcast head-ends, which may be classified as "terrestrial transmitters" for the reception in mainly urban environments, "personal gap fillers" for local/in-door enhancement of satellite signals, and "mobile transmitters" as a complementary moving infrastructure such as on trains or ships.

Both the SC and the CGC are fed by a central Service and Network Head-End (SNHE), which bundles different types of content (TV, IP services, etc.) into a multiplex for transmission over some Distribution Network (DN). The distribution network delivers the multiplex to the satellite broadcast head-end and the CGC, using broadcast infrastructure such as DVB-S2, fibre, or xDSL. All transmitters (modulators) may be operated in Single Frequency Network (SFN) configuration, which improves reception performance, supports seamless handover between transmitters, and avoids the bandwidth overhead inherent to traditional multi-frequency network (MFN) planning.

3   Transmission over the DVB-SH Distribution Network

The DVB-SH [2] specification states that the SNHE is responsible for encapsulation of the video streams and other input data into a constant bit-rate transmission multiplex consisting of MPEG-2 TS or GSE streams. This works well for traditional MPEG-2 TS but raises a number of issues for the GSE protocol, as discussed below.

3.1   GSE-Based Delivery

Direct transmission of a GSE multiplex is (only) possible when a second-generation DVB link (e.g., DVB-S2) is utilized as a DN. However, this has several implications. Retransmission of the multiplex at the broadcast head-ends (SC and CGC) requires GSE packet refragmentation because the GSE protocol is dependent on the underlying base-band frame format, which is different in the DVB-SH standard than in other second-generation DVB specifications. This means that a constant bit rate from the SNHE to the receivers cannot be guaranteed. In MPEG-2 TS systems, this is required to achieve a constant end-to-end delay, allowing receivers to synchronize with the source based on Program Clock Reference (PCR) values in the multiplex. This synchronization model is not required for GSE-based video transmission because synchronization can be effected by a receiver buffering model and the use of timestamps in the RTP protocol, used for video delivery over IP streams. However, some additional overhead at the DVB-SH link may need to be accounted for due to the different GSE encapsulation. Delivery of a transmission multiplex implies a static mapping from DN Input Streams or radio channels to DVB-SH Input Streams or channels. In this case, signalling information carried within the multiplex is destined for the DVB-SH link and must not be interpreted by the DN receiver equipment.

The use of time-slicing such as defined in the DVB Data Broadcasting specification within GSE streams requires knowledge about the bit-exact transmission over the DVB-SH link. This is a caveat for SNHEs delivering a GSE multiplex because the DVB-SH link encapsulation (GSE framing, base-band framing, and physical-layer signalling) needs to be carried out twice, once logically at the SNHE, and once at the set of transmitters. The requirement for knowledge about SH link encapsulation also applies to MPEG-2 TS transmission - in the DVB-SH Implementation Guidelines, a (suboptimal) solution is described based on calculating the approximate starts of time-slices assuming an average constant bit-rate. In place of the SNHE, the broadcast head-ends could be responsible for inserting or updating time-slicing information in the GSE streams. This would add unwanted complexity to transmitters acting as pure modulators.

In SFN operation, the SNHE must deliver synchronization information along with the multiplex to the set of transmitters within the DVB-SH network. For MPEG-2 TS transmission, an SH Frame Initialization Packet (SHIP) carried within a single TS cell has been specified in the SH standard. As the SHIP specifies all the physical-layer settings and the beginning of SH Frames, the start of each EFrame can be computed. Thus, an adaptation of the SHIP to the GSE protocol could support SFN operation provided that refragmentation is carried out in the same way at all transmitters and the possible additional refragmentation overhead is accounted for at the SNHE. Note that in difference to time-slicing an approximate calculation of synchronization time-stamps based on an assumed average bit-rate cannot be used. Furthermore, even when bit-exact broadcast transmission is replicated at the SNHE, loss of a single GSE packet at a modulator affects an entire SH Frame because the exact number of bytes to be left out due to GSE refragmentation cannot be derived. This means that the transmitter cannot stay in synchronization with the other broadcast head-ends for the rest of the SH Frame.

3.2   IP-Based Delivery

The DN may be IP-based. In this case, the SNHE and the transmitters are end nodes in an IP infrastructure, and the connection may consist of several intermediate links, possibly over different physical media. This requires the carriage of GSE streams over some IP-based tunnelling mechanism. While this is possible and can avoid the need for GSE refragmentation at the broadcast head ends as long as packets are reliably transmitted or their offset within base-band frames is known, base-band framing and physical-layer signalling still needs to be considered at SNHEs while being carried out at transmitters.

3.3   Transmission of Base-Band Frames

The DVB-T2 [3] standard defines a specific Modulator Interface Protocol (MIP) [4] for the carriage of base-band frames and physical-layer signalling over MPEG-2 TS or IP links. Delivering base-band frames (containing GSE data) instead of raw GSE packets has a number of advantages: It allows control over each frame, such as over the Input Stream Identifier (ISI) and the use of a base-band frame CRC [5], physical-layer settings and other transmitter instructions can be directly signalled instead of indicated indirectly via some GSE extension header (e.g., a GSE adaptation of the SHIP [5]), and loss of a base-band frame does not affect the entire SH Frame in an SFN setting. Moreover, it does not require broadcast head-ends to carry out any further encapsulation.

4   Modulator Interface Protocol

Because of the advantages of an MIP compared to the delivery of raw GSE streams such a protocol, applicable to DVB-SH systems, is introduced in this section. The MIP is also suggested when the DN consists of a DVB second-generation broadcast link because of the little overhead it introduces. In the case that the broadcast transmission of another DVB system is to be forwarded over DVB-SH (terrestrial) components/repeaters it is recommended that a gateway be defined that is acting as an SNHE and carries out the re-encapsulation of the GSE streams present in the multiplex.

A number of requirements can be identified for the MIP:

− Support for the carriage of EFrames and physical-layer signalling information
− Support for the delivery of multiple transmission multiplexes over a single link
− In-order delivery of data and fragments thereof
− Reliable detection of packet errors and of missing packets
− Identification of the SH Frame associated with an EFrame or physical-layer signalling packet
− Identification of the relative position of an EFrame within a particular SH Frame
− Transparent carriage over different transport protocols, including UDP, TCP, and RTP

The protocol design, depicted in Figure 1, has been chosen to fulfil these requirements. Each MI packet is made up of a 3-byte header, a variable-length payload field, and a CRC-32. The header includes a 4-bit *version* identifier to consider future changes to the protocol, a 12-bit *stream identifier*, which allows for the parallel transport of independent multiplexes, and an 8-bit *payload type* field that indicates the format of the payload field. The *version* field is set to zero for this specification of the protocol. The *stream identifier* may be chosen by the MI encapsulator or negotiated out-band between encapsulator and the modulator side. Physical-layer settings pertaining to a stream may be signalled

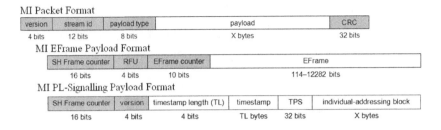

Fig. 1. MI Packet Format

in-band within the protocol or defined externally (e.g., via service-level agreements). The *CRC* that follows the payload is intended for the reliable detection
of bit errors in transmission data or signalling information, and further considers
the quasi error free (QEF) requirements of the GSE protocol.

Two *payload type* values are defined, one for the transmission of EFrames and
one to configure the physical-layer settings of the DVB-SH carrier on a per-SH
Frame basis. An MI packet carrying EFrame data is indicated by a type code of
value 0x00. Its payload field (see Figure 1) contains a 16-bit *SH Frame counter*, 4
reserved (*RFU*) bits, a 10-bit EFrame counter, and the EFrame data, which includes the 114-bit header and the Data Field. EFrame padding may be omitted
from the payload, though zero-bits must be used to align the Data Field to a multiple of 8 bits. The *EFrame counter* is reset to zero for the first EFrame within an
SH Frame and incremented for each subsequent frame. The *SH Frame counter* is
incremented for each SH Frame within a multiplex having the same *stream identifier*; its initial value in the distribution of the stream is unspecified. Both coun-
ters together allow the correct assignment of physical-layer settings per SH Frame
and the reliable detection of lost packets in SFN scenarios. In MFN configuration,
when constant bit-rate transmission and end-to-end delay is required (e.g., for rate
control at the transmitters, or for backwards compatibility with MPEG-2 TS receivers when some EFrames contain MPEG-2 TS data) modulators can derive the
constant number of EFrames per SH Frame via the physical-layer information signalled per SH Frame (described below).

MI packets for the signalling of physical-layer parameters are indicated by a
payload type of value 0x01. The payload field of these packets includes a 16-bit
SH Frame counter, which indicates the SH Frame to which the data applies,
4-bit *version* information, as well as signalling carried in the SH Frame Initialization (SHIP) packet defined for SFN configurations in the DVB-SH standard.
This data is inserted once per SH Frame, and should precede any MI packets
containing EFrame payload data for that frame. The exact format of the payload structure is shown in figure 1. The *timestamp length* field indicates the
length (and format) of the *timestamp* field, and may be set to zero to omit the
timestamp information in MFN signalling. A variable-length *timestamp* field
has also been chosen to support transmission delays in the distribution network
that are larger than 1 second, which is the currently permitted upper limit for

SFN signalling via the SHIP. The support for larger delays is crucial for distribution via non-dedicated IP network connections. The maximum delay field of the SHIP has been omitted as the encapsulator can adjust the *timestamp* values accordingly by referring to the frame emission times at the modulators. The *TPS* (Transmission Parameter Signalling) contains the physical-layer signalling for OFDM transmitters in the DVB-SH system. Signalling for a TDM satellite head-end can be delivered in a transmitter-specific function encoded in the *individual-addressing block*, which is composed of the individual-addressing fields described in the SHIP (i.e., all bytes following and including the *individual_addressing_length* field).

The MI protocol may be delivered over a variety of transport protocols (see Figure 2). Typical solutions include the transport over UDP and RTP, which can provide certain timeliness guarantees. If reliable transmission is desired TCP may be used as a transport protocol. In this case, an adaptation layer in the form of a 2-byte length field preceding each MI packet is needed. Such a solution is compatible to the approach specified for RTP delivery over TCP.

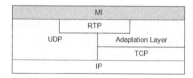

Fig. 2. MI Protocol Stack

5   Conclusion

Convergence to IP networks has become an essential functionality in the architecture of DVB networks. In this contribution we discussed the need for a Modulator Interface Protocol to enable the seamless transport of IP data over the DVB-SH distribution network and presented such a protocol. It supports the carriage of EFrames as well as physical-layer signalling for MFN and SFN transmission, and can be flexibly forwarded over DVB and IP links using various transport protocols including UDP, TCP, and RTP.

References

1. ETSI TS 102 606, Digital Video Broadcasting (DVB); Generic Stream Encapsulation (GSE) Protocol (July 2007)
2. ETSI EN 302 583 V1.1.1 (2008-03) Digital Video Broadcasting (DVB); Framing Structure, channel coding and modulation for Satellite Services to Handheld devices (SH) below 3 GHz
3. ETSI TS 302 755 v1.1.1, Framing structure, channel coding and modulation for a second generation digital terrestrial television broadcasting system (DVB-T2) (September 2009)
4. ETSI TS 102 773 v1.1.1, Digital Video Broadcasting (DVB); Modulator Interface (T2-MI) for a second generation digital terrestrial television broadcasting system (DVB-T2) (September 2009)
5. Study of Generic Stream Encapsulation (GSE) Application to the DVB-Family Standards, ESA study ESA22471/09/NL/AD (April 2011)

Two-Dimensional Markov Chain Model for Performance Analysis of Call Admission Control Algorithm in Heterogeneous Wireless Networks

Sha Sha, Rosemary Halliwell, and Prashant Pillai

School of Engineering, Design & Technology
University of Bradford, Bradford, West Yorkshire, BD7 1DP, United Kindom
{ssha}@student.bradford.ac.uk,
{r.a.halliwell,p.pillai}@bradford.ac.uk

Abstract. This paper proposes a novel call admission control (CAC) algorithm and develops a two-dimensional markov chain processes (MCP) analytical model to evaluate its performance for heterogeneous wireless network. Within the context of this paper, a hybrid UMTS-WLAN network is investigated. The designed threshold-based CAC algorithm is launched basing on the user's classification and channel allocation policy. In this approach, channels are assigned dynamically in accordance with user class differentiation. The two-dimensional MCP mathematical analytic method reflects the system performance by appraising the dropping likelihood of handover traffics. The results show that the new CAC algorithm increases the admission probability of handover traffics, while guarantees the system quality of service (QoS) requirement.

Keywords: Call admission control, handover, two-dimension markov chain, heterogeneous networks, dropping probability.

1 Introduction

In the last years, there is an increasing attention towards the transmission of multimedia applications and services over heterogeneous wireless networking technologies [5]. With the increasing demand for mobile multimedia service, the next generation wireless networks are expected to eventually combine multiple radio technologies [1], purvey the high throughput IP-connectivity to users and achieve service roaming across integrated radio access technologies (RATs). However, this hybrid network architecture requires many technical challenges and functions, including seamless mobility, vertical handovers between diverse RATs, security, subscriber administration, quality of service (QoS) and service provisioning [4].

When a mobile subscriber moves across the overlap networks during its lifetime, handover procedures will be delivered. Meanwhile, new connection requests issued by the intrinsic users are coming forth.

The phenomenon is visualized that handover traffics and new connection will scramble for available radio resource of the target network. That may primarily cause

P. Pillai, R. Shorey, and E. Ferro (Eds.): PSATS 2012, LNICST 52, pp. 180–188, 2013.

the handover dropping by virtue of the limitation of wireless resource and the dynamics large number of users' requests. It is apparent that the increasing of customs and subscriber's mobility lead to the predicament of scarcity limited radio resource allocation and QoS degradation. Hence, QoS provision is an increasingly important task in next generation integrated networks.

One of the key elements in providing QoS guarantees is an effective call admission control (CAC) policy, which not only ensures that the network meets the QoS requirements for new coming traffics but also guarantees that the QoS of the existing does not deteriorate [1]. So it is a tendency to develop an evolved CAC policy for the intricate environment and requirements for differentiating services. This paper is going to present a threshold-based CAC, which sorts users into different classes and uses two-dimensional markov chain process (MCP) to analyze system performance.

The reminder of this paper is organized as follows: Section 2 enumerates CAC for homogeneous and heterogeneous networks respectively; in Section 3, a two-dimension analytical model is revealed; Section 4 presents the numerical results to discuss the performance of new CAC algorithm; finally, this paper is concluded in Section 5.

2   Related Literature

Resource allocation schemes dealing with homogeneous networks have been devised and studied [1]; while a serial of revised CAC algorithms for heterogeneous wireless network have been investigated for a long time. In this section, several proposed works related to the CAC strategies will be introduced for both homogeneous and heterogeneous networks.

2.1   Admission Control Algorithms in Homogeneous Networks

In [3], dynamic channel reservation scheme (DCRS) allows assigning the guard channels reserved for handover traffics to new connection services basing on the request probability to increased channel utilization [2]. It keeps the new connection service blocking probability as low as possible but only provides acceptable quality of handoff services [3].

For handover traffics, dual threshold bandwidth reservation (DTBR) accomplishes the maximum efficiency and maintains the other relative call blocking probability [10]. It employs two thresholds to dominate new connection request voice traffics and data service that involves both handover and new connection request data traffics [7].

2.2   Admission Control Algorithms in Heterogeneous Networks

A great of deal of resource allocation mechanisms are previously addressed for the integrated environment. Ramjee et al. proved Guard channel scheme (GCS), which provides the reserving channels for handover traffic to show its priority. It implements low dropping probability for handover services, but increases new connection services blocking likelihood and degrades the resource utilization [11]. In [8], a CAC algorithm is nominated for voice and data traffics in integrated cellular and WLAN networks. [9] gives the highest priority to the sensitive traffics and degrades the lowest priority connection according to per class degradation[7]. And [12] improves GCS by using a two-dimensional stochastic process model.

3   Analytical Framework for Admission Control Scheme

3.1   Integrated UMTS-WLAN Network

The combination between third-generation (3G) cellular and the IEEE 802.11/16 based wireless networks has been considered as a suitable and viable evolution path toward the next generation of wireless networks [6].

This paper focuses attention on a single UMTS cell and two WLANs, which is shown in Figure 1. In this paper, three classes of user are defined according to the user's moving tendency. *User 1* previously connects with *WLAN A* and tends to move towards to *WLAN B* passing through overlap area. *User 2* is served by *UMTS* network and going to enter *WLAN B* coverage are. *User 3* stays in a stable situation and always has connection with *WLAN B*.

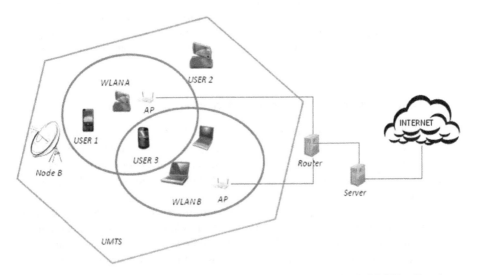

Fig. 1. This figure describes an integrated network consisting of one single UMTS cell and two WLANs. There are three types of user, which are classified according to their mobility.

User 1 and *User 2* send up handover traffics; *User 3* sets up new connection services. Generally speaking, customers prefer to give a higher priority to handover traffics rather than new connection services, especially an ongoing handover traffic. Handover may bring service break off, and it is more annoying to have a call abruptly terminated in the duration of the connection than being blocked occasionally on a new connection attempt [13].

On the other hand, *User 1* ought to bear a higher predominance than *User 2*. The reason is that when *User 1* moves out of *WLAN A* towards to *WLAN B*, *WLAN A* will release the connection with *User 1*, once *User 1* reaches the coverage area of *WLAN B*. If *WLAN B* does not assign the radio resource to *User 1*'s handover traffic, that will cause disruption; while *User 2* has the capability of keeping a continuous serving by *UMTS* network, even if *WLAN B* rejects to permit the channel request from *User 2*. *User 1* is claimed precedence over *User 2* for avoiding service termination. Hence, the channel requests from *User 1* are treated as the highest priority; *User 2* has the intermediate and the resource request from new connection has the lowest priority.

Assuming that the target *WLAN B* has a total channel capacity *C* units and each radio request will occupy one unit. A new connection will be accepted if the occupied channels do not reach *Threshold 1* and a low-priority handover traffic is served when the amount of used resource is not up to *Threshold 2*. A high-priority handover traffic will be accepted as long as there are free channels. Figure 2 shows the channel allocation theory of this new CAC scheme for handover traffics and new connection services.

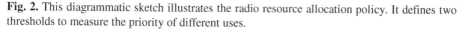

Fig. 2. This diagrammatic sketch illustrates the radio resource allocation policy. It defines two thresholds to measure the priority of different uses.

This threshold-based CAC reflects the priority of each kind of user. A new resource request comes, if there are Th_1 channels are used, the resource request from *User 3* will be denied; once the used channels are up to the amount of Th_2, in that case, not only *User 3*'s but also *User 2*'s channel requests are rejected; with the growth of served traffics, the available radio resource go critical and come of *C*, by then, any channel request will be dropped, even sent from *User 1*.

3.2 Two-Dimensional Traffic Model

Basing on the expatiation above, a two-dimensional MCP system is used to model the prioritized- based CAC algorithm and analyze the performance. The corresponding markov state diagram is portrayed in Figure 3.

Let v_1, v_2 and v_3 are channel request rates of high priority handover traffic, low priority handover traffic and new connection service; the mean serving times for them are $1/\xi_{HH}$, $1/\xi_{LH}$ and $1/\xi_N$, which are following a negative exponential distribution.

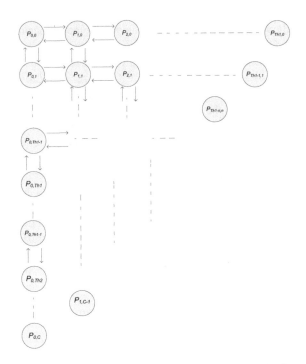

Fig. 3. This is two-dimensional Markov Chain Model. $P_{x,y}$ stands for the steady state probability. Defining Th_1 and Th_2 are thresholds for this scenario.

The intensity of channel request is expressed:

$$I = \frac{v_H}{\xi_H} + \frac{v_3}{\xi_N} \tag{1}$$

Where v_H and $1/\xi_H$ are average channel request rate and service time of handover traffics (both high and low priority).

In this proposed scheme, the performance evaluated parameters are handover traffic dropping probability and new connection blocking probability. The possible state spaces are depicted:

$$S = \{(x,y)|\ x+y \leq C\}. \tag{2}$$

Two-dimensional model settles one server for this system. Each state (x, y) demonstrates the amount of occupied channels: the value of x represents channel number occupied by new connection service and y specifies the quantity of used channel by (high/low priority) handover traffics. Let $S(x,y;x',y')$ stands for the transition rate from state (x,y) to state (x',y') [2]. $P_{x,y}$ clarifies the steady state probability. The channel request rates of new connection and handover traffics are assumed to follow a Poisson arrival process.

Hence, the equation of $S(x,y;x',y')$ is denoted:

$$
\begin{cases}
S(x,y;x,y+1) = v_H, & (0 < x \le Th_1, x+y < Th_1) \\
S(x,y;x,y-1) = y\xi_H, & (0 < x \le Th_1, x+y < Th_1) \\
S(x,y;x+1,y) = v_3, & (0 < x \le Th_1, x+y < Th_1) \\
S(x,y;x-1,y) = x\xi_N, & (0 < x \le Th_1, x+y < Th_1) \\
S(x,y;x,y+1) = v_H, & (0 < x \le Th_1, Th_1 \le x+y < Th_2) \\
S(x,y;x,y-1) = y\xi_H, & (0 < x \le Th_1, Th_1 \le x+y < Th_2) \\
S(x,y;x,y+1) = v_1, & (0 < x \le Th_1, Th_2 \le x+y < C) \\
S(x,y;x,y-1) = y\xi_{HH}, & (0 < x \le Th_1, Th_2 \le x+y < C)
\end{cases}
\tag{3}
$$

The equation of state probability $P_{x,y}$ is exhibited as follows:

$$
P_{x,y} =
\begin{cases}
\dfrac{P_{0,0}}{x!y!}\left(\dfrac{v_3}{\xi_N}\right)^x \cdot \left(\dfrac{v_H}{\xi_H}\right)^y, & \text{where } 0 \le x \le Th_1, x+y \le Th_1 \\[2ex]
\dfrac{P_{0,0}}{x!y!}\left(\dfrac{v_3}{\xi_N}\right)^x \cdot \left(\dfrac{v_H}{\xi_H}\right)^{Th_1-x}\cdot\left(\dfrac{v_H}{\xi_H}\right)^{x+y-Th_1}, & \text{where } 0 \le x \le Th_1, Th_1 \le x+y \le Th_2 \\[2ex]
\dfrac{P_{0,0}}{x!y!}\left(\dfrac{v_3}{\xi_N}\right)^x \cdot \left(\dfrac{v_H}{\xi_H}\right)^{Th_1-x}\cdot\left(\dfrac{v_H}{\xi_H}\right)^{Th_2-Th_1}\cdot\left(\dfrac{v_1}{\xi_{HH}}\right)^{x+y-Th_2}, & \text{where } 0 \le x \le Th_1, Th_2 \le x+y \le C
\end{cases}
\tag{4}
$$

$P_{0,0}$ is the steady state probability of the system being idle[2]. According to the normalization equation $\Sigma_{x,y}P_{x,y}=1$, $P_{0,0}$ is obtained:

$$
P_{0,0} = \Bigg[\sum_{x=0}^{Th_1}\frac{1}{x!}\left(\frac{v_3}{\xi_N}\right)^x \cdot \sum_{y=0}^{Th_1-x}\frac{1}{y!}\left(\frac{v_H}{\xi_H}\right)^y
$$
$$
+ \sum_{x=0}^{Th_1}\frac{1}{x!}\left(\frac{v_3}{\xi_N}\right)^x\left(\frac{v_H}{\xi_H}\right)^{Th_1-x} \cdot \sum_{y=Th_1-x+1}^{Th_2-x}\frac{1}{y!}\left(\frac{v_H}{\xi_H}\right)^{x+y-Th_1}
\tag{5}
$$
$$
+ \sum_{x=0}^{Th_1}\frac{1}{x!}\left(\frac{v_3}{\xi_N}\right)^x\left(\frac{v_H}{\xi_H}\right)^{Th_1-x}\left(\frac{v_H}{\xi_H}\right)^{Th_2-Th_1} \cdot \sum_{y=Th_2-x+1}^{C-x}\frac{1}{y!}\left(\frac{v_1}{\xi_{HH}}\right)^{x+y-Th_2} \Bigg]^{-1}
$$

Recall that once there are no free channels, high priority handover traffics are dropped, thus dropping probability P_{HH} is achieved:

$$
P_{HH} = \sum_{x=0}^{Th_1}\frac{P_{0,0}}{x!(C-x)!}\left(\frac{v_3}{\xi_N}\right)^x\cdot\left(\frac{v_H}{\xi_H}\right)^{Th_1-x}\cdot\left(\frac{v_H}{\xi_H}\right)^{Th_2-Th_1}\cdot\left(\frac{v_1}{\xi_{HH}}\right)^{C-Th_2}
\tag{6}
$$

When the amount of occupied channels are same as Th_2, the handover traffics from low priority class of user will not gain the services and be dropped, hence, the dropping probability of low priority handover traffics is depicted as formula:

$$P_{LH} = \sum_{x=0}^{Th_1} \frac{P_{0,0}}{x!(Th_2 - x)!} \cdot \left(\frac{v_3}{\xi_N}\right)^x \cdot \left(\frac{v_H}{\xi_H}\right)^{Th_1-x} \cdot \left(\frac{v_H}{\xi_H}\right)^{Th_2-Th_1} +$$

$$\sum_{x=0}^{Th_1} \frac{P_{0,0}}{x!} \left(\frac{v_3}{\xi_N}\right)^x \left(\frac{v_H}{\xi_H}\right)^{Th_1-x} \left(\frac{v_H}{\xi_H}\right)^{Th_2-Th_1} \cdot \sum_{y=Th_2-x+1}^{C-x} \frac{1}{y!} \left(\frac{v_1}{\xi_{HH}}\right)^{x+y-Th_2}$$

(7)

Therefore, the total dropping probability of handover traffic is the sum of P_{HH} and P_{LH}.

The expression of the system utilization is profiled as the ratio of the used channels to the whole channel capacity [7]:

$$U = \frac{\sum_{x,y} x y P_{x,y}}{C}$$

(8)

4   Mathematical Results

The handover dropping probability is a key measurement of evaluating the system QoS. Thus, in this section, numerical results will be shown in Figure 4 and Figure 5.

In order to analyze the performance of class-based CAC approach, CAC without threshold and with one threshold schemes are introduced: no threshold scheme assigns all available channels to handover and new connection traffics coequally; one threshold method considers low priority handover traffics have the same level of new connection requests.

Assuming that the total capacity of available channel $C=50$, $1/\xi_{HH}=1/\xi_{LH}=1/\xi_N=150s$, $v_1=0.2\sim0.9$ channel/s, $v_2=0.25\sim1$ channel/s and $v_3=0.25\sim0.8$channel/s.

Figure 4 shows that the dropping probability of handover traffics in CAC without and with one threshold strategies. The horizontal axis stands for the handover intensity and the vertical axis represents the dropping probability of high priority handover traffics. With the increasing of traffics intensity, the dropping probability is also elevating. It is obvious that the CAC without threshold has a higher dropping probability then that of one threshold scheme.

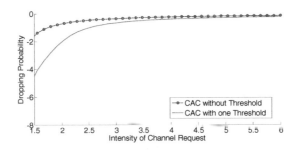

Fig. 4. It plots the dropping likelihood curves for high priority handover traffics of no threshold CAC and one-threshold CAC strategies

Figure 5 explores the dropping probability of high priority handover traffic in one- and two-threshold schemes. Two-threshold based CAC provides a dual-guard for high priority handover traffics; more channels are provided to the highest priority traffics. Therefore, two-threshold method permits more handover traffics to obtain radio channels than other two types of services.

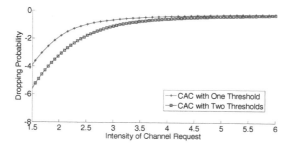

Fig. 5. It appears that two-threshold approach have an advantage in protecting the QoS of high priority handover traffics

The numerical results explicit that two-threshold strategy produces a better performance than no-threshold CAC and one-threshold CAC method in the matter of cutting down the handover services dropping probability and improve the quality of handover services [7].

5   Conclusions

To sum up, this paper utilizes two-dimensional MCP model to resolve class-based CAC algorithm for the next generation wireless networks. This novel method classifies users into distinct levels and assesses the system performance by comparing dropping probability of high priority handover traffics. This approach decreases the dropping probability of handover service, minimizes the dropping likelihood of the user with the highest priority and guarantees the quality of transferred traffic during its lifetime [7].

References

1. Andrews, N., Kondareddy, Y.R., Agrawal, P.: Channel Management in Collocated WiFi-WiMAX Networks. 42nd IEEE Southeastern Symposium on System Theory 2010 , Tyler, Texas(2010)
2. Candan, I., Salamah, M.: Analytical Modeling of a Time- Threshold Based Bandwidth Allocation Scheme for Cellular Networks. Computer Communications (2006)
3. Kim, Y.C., Lee, D.E., Lee, B.J., Kim, Y.S., Mukherjee, B.: Dynamic Channel Reservation Based on Mobility in Wireless ATM Networks. IEEE Communications Magazine, vol. 37, no. 11, pp. 47–51 (1999)

4. Politis, I., Dagiuklas, T., Tsagkaropoulos, M., Kotsopoulos, S.: Interworking Architectures of 3G and WLAN towards All-IP Architectures: Comparisons. Encyclopedia of Mobile Computing and Commerce Vol. 1 Idea Group Inc, accepted for publication (2006)
5. Politis, I., Tsagkaropoulos, M., Dagiuklas, T., Kotsopoulos, S.: Study of the QoS of Video Traffic over Integrated 3G-WLAN systems. 2nd International Mobile Multimedia Communications Conference, Alghero - Sardinia, Italy, September 18-20 (2006)
6. Salkintzis, A.K., Dimitriadis, G., Skyrianoglou, D., Passas, N., Pavlidou, N.: Seamless Continuity of Real-Time Video Across UMTS and WLAN Networks: Challenges and Performance Evaluation. IEEE Wireless Communications (2005) 8–18(2005)
7. Sha, S., Halliwell, R.: Performance Modeling and Analysis of a Handover and Class-Based Call Admission Control Algorithm for Heterogeneous Wireless Networks. 27th Annual UK Performance Engineering Workshop, July 2011(2011)
8. Song, W., Jiang, H., Zhuang, W., Shen, X.: Resource Management for QoS Support in Cellular/WLAN Interworking. IEEE Network, vol.19, no.5, pp.12-18, Sept-Oct (2005)
9. Wang, X. G., Min, G., Mellor, J. E., Al-Begain, K.: A QoS Based Bandwidth Management Scheme in Heterogeneous Wireless Networks. International Journal of Simulation Systems, Science and Technology, pp. 9-17 (2004)
10. Wu, H., Li, L., Li, B., Yin, L., Chlamtac, I., Li, B.: On Handoff Performance for an Integrated Voice/Data Cellular System. Proc. IEEE Int'l Symp. Personal Indoor and Mobile Radio Comm., vol. 5, pp. 2180-2184, Sept. (2002)
11. Wu, S., Wong, K.Y.M., Li, B.: A Dynamic Call Admission Policy with Precision QoS Guarantee using Stochastic Control for Mobile Wireless Networks. IEEE/ACM Trans. Networking, vol. 10, pp. 257 – 271 (2002)
12. Xhafa, A.E., Tonguz, O.K.: Handover Performance of Priority Schemes in Cellular Networks. IEEE Trans. Vehicular Technology, vol. 57, pp. 565--577 (2008)
13. Ye, J., Shen, X.M., Mark, J.W.: Call Admission Control in Wideband CDMA Cellular Networks by Using Fuzzy Logic. IEEE Transactions on Mobile Computing, v.4 n.2, p.129-141, March (2005)

Providing Authentication in Delay/Disruption Tolerant Networking (DTN) Environment

Enyenihi Johnson, Haitham Cruickshank, and Zhili Sun

Centre for Communication Systems Research (CCSR),
University of Surrey, Guildford, United Kingdom
{e.johnson,h.cruickshank,z.sun}@surrey.ac.uk

Abstract. DTN environment is characterized by intermittent connectivity, high/variable delay, heterogeneity, high error rate and asymmetric data rate amongst others. These characteristics accounts for the poor behavior of Internet protocols in this environment. To address these problems, DTN was conceived and designed together with specialized protocols to carry out its services. Its emergence called for a new concept in security that was considered at the design stage. The main aim of this paper is to propose a traditional cryptography based authentication scheme that does not depend on network administrator's availability during post network authentication communication and facilitates bundle processing by the recipient in the absence of connectivity. In this paper, we present and discuss the system model, the proposed credential and the propose authentication scheme. A simulation framework is developed for the implementation of the proposed and referenced schemes. From the simulation results, the proposed scheme was observed to be independent of network administrator's availability during post network authentication communication and facilitates bundle processing in the absence of connectivity.

Keywords: Security, Delay/Disruption Tolerant Networking (DTN), Authentication, Communication Satellite, Traditional Cryptography (TC), Public Key Infrastructure (PKI).

1 Introduction

Delay/Disruption Tolerant Networking (DTN) [1-4] is a networking architecture designed based on message switching mechanism to provide reliable communication in networking environments with long/variable delay, intermittent connectivity, high packet loss rates, heterogeneity and asymmetric data rate amongst others using store-and-forward operation. The poor behavior of existing internet protocols in DTN led to the design of specialized protocols like Bundle Protocol (BP) to provide DTN services as an overlay network. To facilitate interoperability between heterogeneous networks with different network characteristics with DTN, a new layer called Bundle Layer was introduced between the application layer and the transport layer of the internet protocol stack. The design of DTN and the protocol did not evolve without consideration for security which led to the development of relevant security documentations [5] [6] to address DTN-related security issues. The security

P. Pillai, R. Shorey, and E. Ferro (Eds.): PSATS 2012, LNICST 52, pp. 189–196, 2013.

documentations highlight security requirements, define design considerations, identify possible threats as well as open issues.

From the DTN security documentations and the security analysis in [7], the identified threats this work is designed to address are masquerading, modification and replay. To protect the DTN network from these threats, security solutions are required to support both hop-by-hop and end-to-end services as well as policy based routing. The policy based routing requires that an entity involve in DTN communication must be able to verify the authenticity of both the original sender and intermediate forwarder of the bundles (messages) as well as the integrity of the received bundles. The security blocks required to secure a transmitted bundle are defined and described in [6]. Farrell and Cahill in [7] while highlighting the security issues associated with designing the bundle to contain all the required keys and algorithm(s) for security processing, emphasized the need for an additional authorization checks with PKI as a possible solution. The use of PKI is associated with constraints like authorization server unavailability and limited capabilities of certain nodes for cryptographic operations.

The focus of this paper is to investigate how PKI concept can be used to provide an authentication solution that does not depend on server availability during post trust establishment network communication while taking the capabilities of the entities into consideration. The existing PKI based schemes in DTN are [1] and [8]. These schemes either use certificates and encourage large storage of security credentials or depend on server availability. We propose an authentication scheme that combines both asymmetric and symmetric cryptography to provide source authentication as well as message authentication and integrity. The contributions of this paper are summarized as follows: 1) Implementing traditional PKI to provide trust initiation/establishment; 2) introducing our proposed Authorization Pass (APass) as a substitute for PKI based certificate; 3) proposing a scheme that uses symmetric based Hash-based Message Authentication Code (HMAC) for hop-by-hop bundle authentication and integrity, and asymmetric based APass for source authentication; and 4) evaluating the performance of the proposed and reference schemes through simulation.

2   The System Model

The DTN environment in fig. 1 assigns a communication satellite to a zone with each zone having more than one heterogeneous regional network administered by a regional administrator (gateway). Inter-zonal communication is facilitated through satellite-to-satellite communication. Hierarchical routing is implemented with each satellite maintaining a routing table of all RAs within its zone of location and other satellites in the network. Each RA maintains a routing table containing all RAs within its zone and the designated zonal satellite for the zone. RAs in the common territories between two zones maintain routing tables of RAs in the two zones and the designated satellites. The routing tables are computed from the global topology table generated and updated by DTNNA each time an entity joins the network. The global topology table is accessible whenever DTNNA is online.

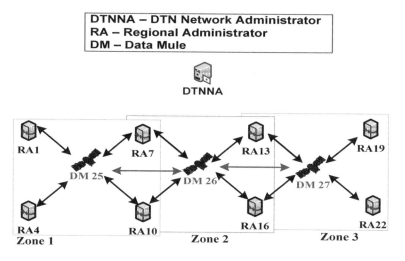

Fig. 1. DTN-Satellite Scenario

The use of satellite is aimed at facilitating bundle transmission between partition or remote networks that cannot communicate directly either due to obstruction or limited range of communication facility. Communication satellites with low computing and storage capabilities are now designed to provide data forwarding services. To minimize cost, Low Earth Orbit (LEO) satellites designed to accommodate limited number of brief contacts at a given time are employed. The limited number of brief contacts at a given time might result in discontinuous coverage. To address this issue, some LEO satellites are designed to provide DTN concept of store and forward services [9]. Security solutions should not be computationally heavy as well as encourage large storage of security credentials.

3   The Proposed APass

The PKI based digital certificate is considered too heavy for implementation in DTN environment. Its usage is associated with revocation and storage of Certificate Revocation List (CRL). It is also identified to provide partial trust management since it does not bind identity to access rights. To address the above issues, we proposed an asymmetric based APass introduced in [10] and shown in fig. 2 which is a modification of the digital certificate. The APass excludes entity's public key considered significantly large and incorporates *role* field to bind entity's identity to authorized action. To offer revocation flexibility and eliminates the storage of CRL, the *validity end* field in the APass is uniform for all APass issued irrespective of when the entities join the network while the *validity start* field differs.

An entity stops sending and receiving bundle when its own APass expires until it gets a new APass from DTNNA since other entities' APass are also assumed to have expired. DTNNA is assumed to know when a new APass must be generated for the entities because allowing the existing APass to expire in the absence of compromise will affect network communication.

```
Authorization Pass
        Authentication Sequence: 1
        Issuer ID: DTNNA@DTN
        Subject ID: RA1@DTN
        Validity Start: 10:00:00 GMT Oct 1 2011
        Validity End: 24:00:00 GMT Dec 31 2011
        Role: A
        Issuer Signature RSAwSHA256: XXXXX
                   XXXXXXXXXXX
```

Fig. 2. The Proposed APass

4   Authentication Model

The authentication model is sub-divided into three phases of registration, network authentication and data exchange. The registration phase is facilitated by a public trusted entity called Registration Authority (RegAuth) which provides security information (secInfo) required for network authentication. The network authentication phase is facilitated by DTNNA and provides credential required for the data exchange phase. The data exchange phase is facilitated by network entities (RAs and DMs) using credentials obtained from DTNNA during network authentication for secure communication. We assume the entities cannot be compromise and the DMs are part of the DTNNA service providing network. Every RA is customized with pre-install initial public/private key pair, device identifier (devID) and RegAuth's public key access on activation. For every node (RA) manufactured the vendor provides RegAuth with node's devID and initial public key. Every node generates its public-private key pair. RAs must pass through the registration/network authentication phase to communicate in the DTN overlay network. RegAuth generates security information required for network authentication, while DTNNA generates trust information for post network authentication communication.

REGISTRATION/NETWORK AUTHENTICATION: The registration and network authentication phases are shown in fig. 3 below.

Registration Phase:
$RA1 \rightarrow RegAuth$: $Pb_{RegAuth}\{rtj\ DTN\ network|devID_{RA1}\}$ (1)
$RegAuth \rightarrow RA1$: $Pb_{RA1}\{devID_{RA1}|\ secInfo_{RA1}|ID_{DTNNA}|secInfo_{DTNNA}|Pb_{DTNNA}\}$ (2a)
$RegAuth \rightarrow DTNNA$: $Pb_{DTNNA}\{nfa|devID_{RA1}|\ secInfo_{RA1}\}$ (2b)

Network Authentication Phase:
$RA1 \rightarrow DTNNA$: $Pb_{DTNNA}\{authReq|devID_{RA1}|\ secInfo_{RA1}|Pb_{RA1}\}$ (3)
$DTNNA \rightarrow RA1$: $Pb_{RA1}\{authConf|ID_{DTNNA}|secInfo_{DTNNA}|K_{dtn}|APass_{RA1}\}$ (4)

RSA encryption/decryption is used to secure communication with entity's public key used to encrypt and private key to decrypt. DTNNA and RA1 use security information from RegAuth for mutual authentication during network authentication where received secInfo must match the one obtained from RegAuth. RA1 uses Pb_{DTNNA} obtained in 2a to encrypt 3 while DTNNA uses Pb_{RA1} obtained in 3 to encrypt 4.

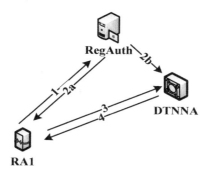

Fig. 3. Registration/Network Authentication Phase

Notations: Pb_{RA1}, Pb_{RA7}, $Pb_{RegAuth}$, Pb_{DTNNA}: Public keys of RA1, RA7, RegAuth and DTNNA; $secInfo_{RA1}$, $secInfo_{DTNNA}$: Security information of RA1 and DTNNA; $APass_{RA1}$, $APass_{DM25}$: APass of RA1 and DM25; $devID_{RA1}$: Device identifier of RA1; K_{dtn}: Network-wide shared symmetric key; N_{RA1}: Nonce generated by RA1(randomly generated string); *Tstmp*: Bundle's timestamp; ID_{DTNNA}, ID_{RA1}, ID_{RA7}: Network identifiers of DTNNA, RA1 and RA7; CustID: Reserved for network identifiers of the data mules; Pb_{RA7}{Hello}: Encrypted payload processed only by RA7; {$APass_{RA1}$| N_{RA1}}: Access control block; {Tstmp | ID_{RA1} | CustID | ID_{RA1}}: Our primary bundle block; |: Concatenated operation; rtj: Request to join; nfa: Notification for authentication; authReq: Authentication request; authConf: Authentication confirmation.

DATA EXCHANGE PHASE: The entities (RAs/DMs) are assumed to have passed through registration/network authentication and in custody of $secInfo_{DTNNA}$ and K_{dtn} for HMAC computation/verification, their respective APass and Pb_{DTNNA} for DTNNA signature verification. The authenticated resource is the bundle and is designed to provide authentication, integrity and confidentiality. Our version of bundle and acknowledgement considering the first hop between RA1 and DM25 with RA7 as destination in fig. 1 is shown below.

(1). RA1 → DM25: (Pb_{RA7}{Hello}|{$APass_{RA1}$| N_{RA1}}|{Tstmp|ID_{RA1} | CustID |ID_{RA7}}
)·hmac

(2). DM25 → RA1: (isAccepted | $APass_{DM25}$ | N_{RA4})·hmac

We assumed the entities are communicating for the first time after network authentication and Pb_{RA7} was obtained by RA1 when DTNNA was online. RA1 generates the bundle in (1), append hmac computed using K_{dtn} and $secInfo_{DTNNA}$ and send to DM25. DM25 upon receiving the bundle verifies the hmac by comparing the appended hmac with the one it computes using K_{dtn} and $secInfo_{DTNNA}$. The bundle content is only access if the appended and computed hmacs matches. DM25 identifies source/destination and ascertains bundle validity and freshness as well as APass signature verification using DTNNA's public key. If all conditions are met, DM25 accepts custody of the bundle and sends acknowledgement (2) appended with hmac to

RA1. DM25 then replaces the content of access control block with its APass (APass$_{DM25}$) and generated nonce (N$_{DM25}$) after which it inserts ID$_{DM25}$ in the CustID field. It then computes hmac and append at the end of the bundle as discussed earlier before forwarding to RA7. RA7 accesses the bundle content after hmac verification to ascertain bundle validity and freshness as well as APass signature verification using DTNNA's public key. If all the conditions are met, it accepts custody and sends acknowledgement containing its APass (APass$_{RA7}$) and the received nonce N$_{DM25}$ DM25. RA7 being the destination then proceeds with the decryption of Pb$_{RA7}${Hello} using its private key. Upon receiving acknowledgement, an entity verifies the appended hmac with the one it computes and verifies received APass without signature verification. The acknowledgement is confirmed as reply to a sent bundle if the received nonce matches the sent nonce after which a stored copy of the sent bundle is deleted. The recipient of a bundle only verifies the last forwarder on the assumption that the last forwarder must have verified the source/previous forwarder before accepting custody of the bundle.

COMPARATIVE ANALYSIS: The proposed scheme is compared with the TC based protocol by Asokan et al in [8] shown in fig. 4 below.

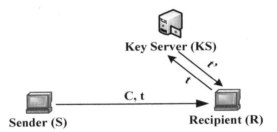

Fig. 4. Compared TC based Authentication Protocol

The protocol assumes that S knows the public key of KS and R identity Id$_R$ prior to communication. KS shares a secret key with each entity and the protocol works as follows: (1) S generates a random secret key (k) and encrypt a message to have C. k and Id$_R$ are then encrypted with KS's public key to have t which is sent together with C to R. (2) R forwards t to KS which decrypts (asymmetric) it with its private key to access the content and verify Id$_R$. KS then encrypts (symmetric) k retrieved from t with the key it shares with R (K$_{SR}$) to have t' forwarded to R. R upon receiving t' decrypts (symmetric) it with K$_{SR}$ to retrieve k used to decrypt (symmetric) C to access the content. This protocol like the proposed scheme is PKI based combining symmetric and asymmetric algorithm and was design for analysis in DTN environment even though it was never validated through simulation. We modelled this protocol and implement it for DTN multi-hop communication. During network authentication phase, DTNNA generates different symmetric keys it shares with each RA/DM instead of K$_{dtn}$ in the proposed scheme. S's APass and DTNNA's APass were included in t and t' to provide source authentication. The access control and primary blocks in the bundle are encrypted with randomly generated k to form C. The random key k is used to encrypt/decrypt the acknowledgement.

5   Simulation and Performance Evaluation

To evaluate the performance of the proposed and compared schemes, the DTN-satellite scenario in fig. 1 was modeled in C++ using Microsoft Visual Studio 2008 with security integrated using Crypto++ Cryptographic Library. The modeled framework implements hierarchical routing with DTNNA the only source of updates. We used Sony VAIO laptop running Windows 7 with the following parameters: Intel CoreTM 2 Duo with T5500 Processor @ 1.66GHz speed, 2.50RAM, 120 GB/Go HDD and a system type of 32 bit Operating System.

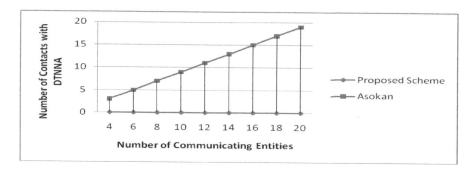

Fig. 5. Simulation result for Number of Contact with DTNNA

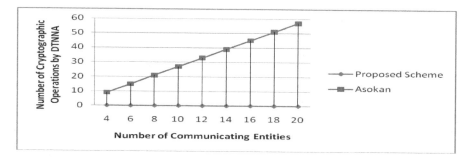

Fig. 6. Simulation result for Number of Cryptographic Operations by DTNNA

Fig. 5 shows the simulation result for the number of contact of the communicating entities with DTNNA while fig. 6 shows the number of cryptographic operations carried out by DTNNA during bundle transmission. The number of communicating entities include one sender, one destination and as many intermediate nodes (DM) as possible. While the proposed scheme has zero contact and zero cryptographic operations by DTNNA, the number of contacts and cryptographic operations by

DTNNA for Asokan increases with increase in the number of communicating entities. For every contact establish DTNNA carries out three cryptographic operations of public key decryption, signature verification and symmetric key encryption.

6   Conclusion

We modeled a PKI-based system model with online server (DTNNA) in C++ using Microsoft Visual Studio 2008 and Crypto++ Cryptographic Library for security integration. We presented and discussed the system model, the proposed APass, proposed authentication model and the TC based protocol in [8] considered for comparative analysis. From the simulation results in the last section, we confirmed that the TC based protocol in [8] depends on server availability and places more load on the server during multi-hop communication. The proposed scheme shows how traditional cryptographic techniques can be used judiciously to provide authentication and integrity solution that facilitates bundle processing by the recipient without depending on server availability. The proposed scheme suits the communication satellite because it facilitates onboard switching and processing in addition to providing lighter cryptographic operations and limited storage. The only issue with our scheme is that the sender will need contact with DTNNA for destination' public key ahead of communication or will have to store the keys.

References

1. Fall, K.: A Delay-Tolerant Network Architecture for Challenged Internets. In: SIGCOMM 2003, Karlsruhe, Germany, August 25-29 (2003)
2. McMahon, A., Farrell, S.: Delay- and Disruption-Tolerant Networking. IEEE Internet Computing 13(6), 82–87 (2009)
3. Cerf, V., et al.: Delay-Tolerant Networking Architecture. IETF Network Working Group RFC 4838 (2007)
4. Scott, K., Burleigh, S.: Bundle Protocol Specification. IETF Network Working Group RFC 5050 (2007)
5. Farrell, S., Symington, S., Weiss, H., Lovell, P.: Delay-Tolerant Networking Security Overview. IETF Internet Draft, draft-irtf-dtnrg-sec-overview-06 (2009)
6. Symington, S., Farrell, S., Weiss, H., Lovell, P.: Bundle Security Protocol Specification. IETF Internet Draft, draft-irtf-dtnrg-bundle-security-15 (2010)
7. Farrell, S., Cahill, V.: Security Considerations in Space and Delay Tolerant Networks. In: Second IEEE International Conference on Space Mission Challenges for Information Technology (2006)
8. Asokan, N., Kostiainen, K., Ginzboorg, J., Ott, J., Luo, C.: Towards Securing Disruption-Tolerant Networking. Nokia Research Centre, NRC-TR-2007-007 (2007)
9. Communication Satellite,
   http://en.wikipedia.org/wiki/Communications_satellite
10. Johnson, H., Cruickshank, H., Sun, Z.: Managing Access Control in Delay-/Disruption Tolerant Networking (DTN) Environment. In: 4th IEEE/IFIP International Conference on New Technology, Mobility and Security (2011)

Author Index

CPSIA information can be obtained at www.ICGtesting.com
Printed in the USA
LVOW011747070713

341741LV00006B/142/P